Vitamins and Minerals Demystified

Demystified Series

Accounting Demystified
Advanced Calculus Demystified
Advanced Physics Demystified
Advanced Statistics Demystified
Algebra Demystified
Alternative Energy Demystified
Anatomy Demystified
asp.net 2.0 Demystified
Astronomy Demystified
Audio Demystified
Biology Demystified
Biotechnology Demystified
Business Calculus Demystified
Business Math Demystified
Business Statistics Demystified
C++ Demystified
Calculus Demystified
Chemistry Demystified
Circuit Analysis Demystified
College Algebra Demystified
Corporate Finance Demystified
Data Structures Demystified
Databases Demystified
Differential Equations Demystified
Digital Electronics Demystified
Earth Science Demystified
Electricity Demystified
Electronics Demystified
Engineering Statistics Demystified
Environmental Science Demystified
Everyday Math Demystified
Fertility Demystified
Financial Planning Demystified
Forensics Demystified
French Demystified
Genetics Demystified
Geometry Demystified
German Demystified
Home Networking Demystified
Investing Demystified
Italian Demystified
Java Demystified
JavaScript Demystified
Lean Six Sigma Demystified
Linear Algebra Demystified

Macroeconomics Demystified
Management Accounting Demystified
Math Proofs Demystified
Math Word Problems Demystified
MATLAB® Demystified
Medical Billing and Coding Demystified
Medical Terminology Demystified
Meteorology Demystified
Microbiology Demystified
Microeconomics Demystified
Nanotechnology Demystified
Nurse Management Demystified
OOP Demystified
Options Demystified
Organic Chemistry Demystified
Personal Computing Demystified
Pharmacology Demystified
Physics Demystified
Physiology Demystified
Pre-Algebra Demystified
Precalculus Demystified
Probability Demystified
Project Management Demystified
Psychology Demystified
Quality Management Demystified
Quantum Mechanics Demystified
Real Estate Math Demystified
Relativity Demystified
Robotics Demystified
Sales Management Demystified
Signals and Systems Demystified
Six Sigma Demystified
Spanish Demystified
SQL Demystified
Statics and Dynamics Demystified
Statistics Demystified
Technical Analysis Demystified
Technical Math Demystified
Trigonometry Demystified
UML Demystified
Visual Basic 2005 Demystified
Visual C# 2005 Demystified
Vitamins and Minerals Demystified
XML Demystified

Vitamins and Minerals Demystified

Dr. Steve Blake

New York   Chicago   San Francisco   Lisbon   London
Madrid   Mexico City   Milan   New Delhi   San Juan
Seoul   Singapore   Sydney   Toronto

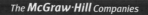

1 2 3 4 5 6 7 8 9 0   DOC/DOC   0 1 3 2 1 0 9 8 7

ISBN   978-0-07-148901-0
MHID   0-07-148901-0

Sponsoring Editor
Judy Bass

Copy Editor
Matthew Kushinka

Production Supervisor
Pamela A. Pelton

Proofreader
D & P Editorial Services, LLC

Editing Supervisor
Stephen M. Smith

Art Director, Cover
Jeff Weeks

Project Manager
Joanna V. Pomeranz

Composition
D & P Editorial Services, LLC

Printed and bound by RR Donnelley.

McGraw-Hill books are available at special quantity discounts to use as premiums and sales promotions, or for use in corporate training programs. For more information, please write to the Director of Special Sales, McGraw-Hill Professional, Two Penn Plaza, New York, NY 10121-2298. Or contact your local bookstore.

CONTENTS

FIGURES, GRAPHS, AND RDA TABLES

FIGURES

GRAPHS

RDA TABLES

PREFACE

This book is for anyone who wants an understanding of the fascinating role vitamins and minerals play in nutrition. It can be used as a supplementary textbook for nutrition classes, as a self-learning guide, and as a refresher for health professionals. This book broadens and explains the vitamin and mineral information found in standard nutrition courses.

Throughout the text are many figures, graphs, and tables that visually display information and relationships. If you have not taken a class in biochemistry, then this will be an interesting and relevant way to be introduced to it.

I recommend that you browse through each chapter or part before studying it in detail. My goal is for you to learn the information easily. You can use this book as a permanent reference as questions on vitamins and minerals come up. There are handy tables in most of the chapters that display the amount of each vitamin and mineral needed by people of different ages. You may find Appendix G to be helpful for reference as it contains the quick summaries for all of the vitamins and minerals. Appendix G has a black tab on each page to make it easy to reference.

There are quizzes at the end of each chapter. These quizzes are meant to be taken "open book" so that you can look up the answers. At the end of each of the four parts there are tests. These tests are meant to be taken "closed book." These questions are designed to help you see which areas you have mastered and which areas need further study. At the end of the book is a final exam of 100 questions. These questions are a bit easier and cover the more important topics. The answers are in the back of the book.

Many vitamins and minerals assist enzymes in building, breaking down, and changing nutrients in our bodies. Enzymes are especially important in releasing the energy contained in food.

Several of the vitamins work as antioxidants to protect us from aging and from chronic diseases. Many of the nutritional minerals also work as essential parts of antioxidant compounds inside our bodies.

When you finish this course, you will have a solid grasp of how vitamins and minerals keep us healthy.

Suggestions for future editions are welcome. Good luck.

Dr. Steve Blake

ACKNOWLEDGMENTS

My deepest gratitude goes to my wife Catherine for her thoughtful editing and encouragement. Special thanks to my sister Carolyn for her support. Thanks to my patient editor, Judy Bass. Thanks go to all those who gave valuable feedback, especially Jim Woessner, M.D., Chris Melitis, N.D., Lori Hager, R.D., Michael Gregor, M.D., Headley Freake, Ph.D., Linda Parker, M.H., Cynthia Peterson, Ph.D., Joanna V. Pomeranz, and Kiki Powers, M.S.

ABOUT THE AUTHOR

Dr. Steve Blake has taught anatomy, physiology, and exercise physiology, and has authored over a dozen major publications including *Alternative Remedies* and *Healing Medicine*. He designed several best-selling vitamins including Children's Chewable and the Advanced Nutritional System, and presented a popular radio show, "Natural Health Tips from Dr. Steve." Dr. Blake has created one of the largest databases of natural remedies from around the world, and developed The Diet Doctor, computer software for evaluating and graphing the nutrients in diets. He has a doctorate in naturopathic medicine as well as a doctorate in holistic health. He and his wife Catherine live in Maui, Hawaii.

PART ONE

The Water-Soluble Vitamins

Introduction to the Water-Soluble Vitamins

Vitamin C and all of the B vitamins are water-soluble vitamins. These water-soluble vitamins circulate freely in the blood, in the watery fluids between cells, and in the fluids inside cells. The solubility of a vitamin affects its mode of action, storage, and toxicity. Most of the water-soluble vitamins can move through the bloodstream without needing carriers for transport; in fact, only vitamin B_{12} needs a binding protein for transport in the bloodstream.

Any excesses of water-soluble vitamins are eliminated by the kidneys. Folates and vitamin B_{12} are exceptions to this rule and are regulated by the liver and released through the bile.

The B vitamins are inactive until they are transformed into their coenzyme form. All of the water-soluble vitamins can act as important parts of the coenzymes that make enzymatic reactions possible, as seen in Figure I-1.

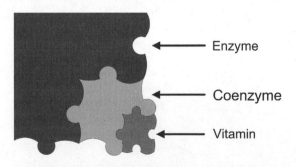

All of the B vitamins and vitamin C can act
as catalysts for enzymatic reactions.

Figure I-1   Vitamins can form part of the
coenzymes that activate enzymes.

It is important to note that water-soluble vitamins are vulnerable to losses during cooking as they can easily leach out into cooking water. Many of these vitamins are sensitive to heat as well. Unfortunately, all types of vitamins are depleted during the refining of grains. With the exception of vitamin B_{12}, these water-soluble vitamins must be eaten regularly as storage in our bodies is limited.

Water-soluble vitamins are not toxic when consumed in food. Supplements of water-soluble vitamins are also not toxic in normal amounts. One exception is supplemental vitamin B_3, but only when taken in the form of niacin, which can cause skin flushing. Another exception is Vitamin C. Vitamin C can cause intestinal irritation, but only when taken in large amounts and in the acidic form (ascorbic acid). When taken in an ascorbated or buffered form, Vitamin C is not irritating, even in large doses.

With the water-soluble vitamins we have the antioxidant support and coenzymes needed to help our systems run smoothly. Read on for more detail about how these wonderful nutrients work.

CHAPTER 1

The B Vitamins
The Energy Vitamins

The B vitamins were identified and isolated early in the twentieth century when refined grains were first found to cause deficiency diseases. The B vitamins work so closely together that it is hard to tell which individual B vitamin is missing when a deficiency occurs. The B vitamins need to be taken together in food or in supplements.

Introducing the B Vitamins

The B Vitamins

B_1 Thiamin	Biotin	Folate
B_2 Riboflavin	B_5 Pantothenic Acid	B_{12} Cobalamin
B_3 Niacin	B_6 Pyridoxine	

Other nutrients also interact with the B vitamins. In one case, a mineral, iron, and an essential amino acid, tryptophan, are both needed to synthesize niacin, vitamin B_3. The best way to avoid a deficiency of B vitamins is to eat a varied diet of fresh fruit, an abundance of vegetables, whole grains, legumes, nuts and seeds, and other food as desired. Some of these B vitamins can also be made by friendly bacteria in a healthy colon and absorbed into circulation.

Metabolism

Catabolism = Breaking down of components
Anabolism = Building up of components

CATABOLISM AND ANABOLISM

The primary role of the B vitamins is catalyzing energy production in the body. One side of metabolism is *catabolism*. Catabolism is the breaking down of carbohydrates, fats, and proteins, often to produce energy, as shown in Figure 1-1. The other side of metabolism is *anabolism*. Anabolism is the building up of components, for example, building a protein from amino acids. The B vitamins are used in many aspects of metabolism, but they are the stars in energy production as coenzymes in catabolic reactions.

A coenzyme attaches to an enzyme to activate the enzyme. These coenzymes enable the enzymes to synthesize compounds or to dismantle compounds.

THE ROLES OF THE B VITAMINS

The B vitamins are needed for healthy nerve conduction and thus muscle action. They are needed for the synthesis of many important neurotransmitters, such as acetylcholine, serotonin, dopamine, and norepinephrine. B vitamins are also indispensable for the synthesis of fats used in the myelin sheaths of nerve cells. With the special ability of the B vitamins to make neurotransmitters and also to make the myelin sheaths of nerve cells, the B vitamins are well known for helping with stress.

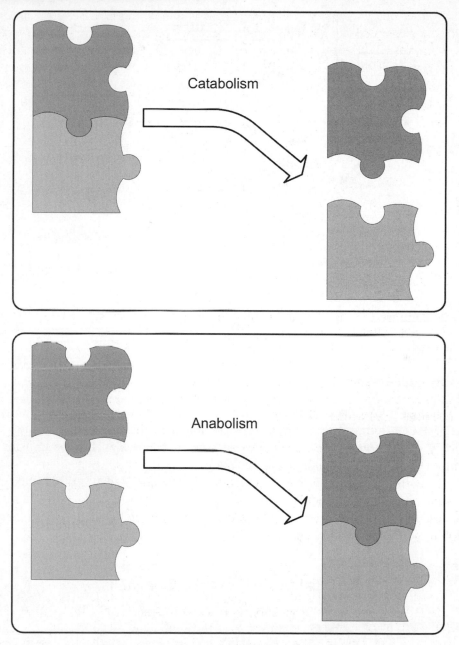

Figure 1-1   The two aspects of metabolism.

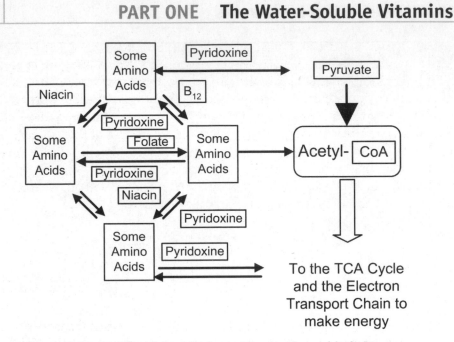

Figure 1-2 B vitamins are used to convert amino acids for energy production.

Another role of the B vitamins is to help us convert the amino acids that make up protein. Amino acids are routinely converted in the cell in preparation for energy production, as seen in Figure 1-2. The B vitamins enable the transfer of amino groups from one amino acid to another.

The ability to transfer amino groups is vital in the maintenance of our DNA. B vitamins help with our metabolism by helping to convert fats and amino acids to glucose (blood sugar). Additionally, certain B vitamins are needed to build hemoglobin, which carries oxygen throughout the body.

Principal Functions of the B Vitamins

Energy production from carbohydrates, fats, and protein
Synthesis of neurotransmitters
Conversion of amino acids
Synthesis of fatty acids and hormones
Antioxidant protection

The vitamin B complex is vital for the synthesis of fatty acids. The B vitamins help us make cholesterol and also help us control cholesterol. They are needed for the synthesis of phospholipids in the all-important cell membranc. They are also needed to synthesize steroid hormones such as melatonin, the sleep hormone.

Some of the B vitamins are useful in protecting us from free radical attack. B vitamins can lower homocysteine levels to reduce our risk of heart disease. They work with important antioxidants such as glutathione. Also, B vitamins help us eliminate certain drugs, carcinogens, and steroid hormones.

These B vitamins are vital to health and life. Now we will take a look at the individual B vitamins. Each one has its own character.

Vitamin B₁—Thiamin, the Carbo Burner

Thiamin was first discovered in Japan in the early 1900s, when thc lack of thiamin in white rice caused *beriberi*. Thiamin was first synthesized in 1936. Thiamin is found in rice bran and rice germ, both of which are removed when white rice is made from brown rice.

Thiamin plays a key role in the metabolism of energy in all cells. Thiamin is part of the coenzyme *ThiaminPyroPhosphate* (TPP), which helps convert pyruvatc to *acetyl-coenzyme A* (also known as *acetyl-CoA*). This is a necessary step in the production of cellular energy from carbohydrates as shown in Figure 1-3. Magnesium is needed to convert thiamin to TPP. Although refined grains are often fortified with thiamin, thcir original magnesium is depleted by an average of 76 percent. Magnesium deficiency may also have played a part in beriberi since so much magnesium is lost when white rice is refined.

Thiamin Coenzyme forms are:
ThiaminPyroPhosphate (TPP)
Thiamin Triphosphate (TTP)

Thiamin is also found in the form of *thiamin triphosphate* in nerve and muscle cells. This form of thiamin activates the transport of electrolytes across the membranes of nerve and muscle cells. This allows healthy nerve conduction and muscle action.

The amount of dietary thiamin needed is based on the amount needed in producing energy. Thiamin needs will be met by most normal diets if enough food is

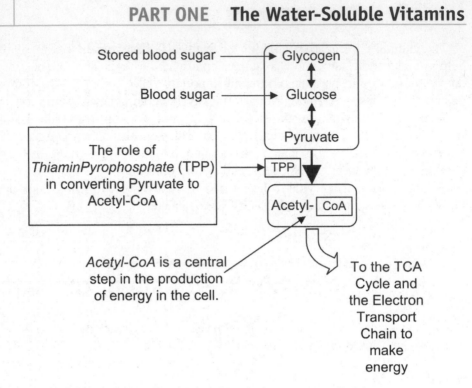

Figure 1-3   The role of thiamin in energy production.

eaten to meet energy requirements. People generally get enough thiamin to meet the *Recommended Daily Allowance* (RDA), although marginal thiamin deficiency affects about one quarter of the people in the United States and Canada. Elderly people are at risk of thiamin deficiency because of low intakes and reduced absorption.

The RDA for men is 1.2 mg per day and for women, 1.1 mg. Please see Table 1-1 for B vitamin RDAs and *Adequate Daily Intakes* (AI) for all ages. When enough information cannot be gathered to establish an RDA, an AI is established as a rough guide. Please refer to Table 1-2 for common supplemental amounts of the B vitamins for adults. These amounts are well above the RDAs, yet still below any established upper intake levels. These amounts may be high enough to compensate for extra stress and food that is depleted of some of its B vitamins.

Thiamin deficiency can result from inadequate food intake. Thiamin deficiency is common among alcoholics, who often have inadequate food intakes. Alcohol provides energy without providing many of the necessary nutrients. Alcohol also impairs the absorption of thiamin, while increasing excretion of thiamin. Enzymes present in raw fish and shellfish destroy thiamin. Also, tannins in tea and coffee can oxidize thiamin, reducing the availability of thiamin in the diet. Extreme thiamin

Table 1-1 RDAs for the B vitamins are **bold** and the AIs are not bold.

B Vitamins RDAs &AIs	Years of Age	B_1 mg	B_2 mg	B_3 mg	Biotin mcg	B_5 mg	B_6 mg	Folate mcg	B_{12} mcg
Infants	0 to ½	0.2	0.3	2	5	1.7	0.1	65	0.4
Infants	½ to 1	0.2	0.4	4	6	1.8	0.3	80	0.5
Children	**1–3**	**0.5**	**0.5**	**6**	8	2	**0.5**	**150**	**0.9**
Children	**4–8**	**0.6**	**0.6**	**8**	12	3	**0.6**	**200**	**1.2**
Boys	**9–13**	**0.9**	**0.9**	**12**	20	4	**1**	**300**	**1.8**
Girls	**9–13**	**0.9**	**0.9**	**12**	20	4	**1**	**300**	**1.8**
Teen boys	**14–18**	**1.2**	**1.3**	**16**	25	5	**1.3**	**400**	**2.4**
Teen girls	**14–18**	**1**	**1**	**14**	25	5	**1.2**	**400**	**2.4**
Men	**< 50**	**1.2**	**1.3**	**16**	30	5	**1.3**	**400**	**2.4**
Women	**< 50**	**1.1**	**1.1**	**14**	30	5	**1.3**	**400**	**2.4**
Men	**> 50**	**1.2**	**1.3**	**16**	30	5	**1.7**	**400**	**2.4**
Women	**> 50**	**1.1**	**1.1**	**14**	30	5	**1.5**	**400**	**2.4**
Pregnancy	**All ages**	**1.4**	**1.4**	**18**	30	6	**1.9**	**600**	**2.6**
Breastfeeding	**All ages**	**1.4**	**1.6**	**17**	35	7	**2**	**500**	**2.8**

Table 1-2 Typical supplement amounts of the B vitamins.

Common Supplement Amounts	B_1 mg	B_2 mg	B_3 mg	Biotin mcg	B_5 mg	B_6 mg	Folate mcg	B_{12} mcg
Adults	50	50	30	60	25	50	800	100

Summary for Thiamin—Vitamin B_1
(Please note that you can find this summary and all of the
summaries in Appendix G in the back of the book.)

Main function: energy metabolism.
RDA: men, 1.2 mg; women, 1.1 mg.
No toxicity reported, no upper intake level set.
Deficiency disease: beriberi.
Healthy food sources: whole grains, and found in most raw
or lightly cooked foods.
Degradation: easily destroyed by heat.
Coenzyme forms: *ThiaminPyroPhosphate* (TPP), *Thiamin
Triphosphate* (TTP).

deficiency can lead to an enlarged heart, weight loss, muscular weakness, poor short-term memory, and cardiac failure.

If thiamin is absent or too low in the diet for prolonged periods, this can result in beriberi. Beriberi can result from the consumption of unfortified refined grains such as white flour. Beriberi can cause damage to the nervous system, heart, and muscles.

Milligrams and Micrograms

One gram   = 1000 mg (milligram)
One milligram = 1000 mcg (microgram)

Some vitamin and mineral amounts are measured in milligrams and some are measured in micrograms.

Thiamin is found in small but sufficient quantities in most nutritious foods, especially whole grains; please refer to Graph 1-1. Only highly refined foods are lacking in thiamin. Pork products are very high in thiamin. Healthy foods rich in thiamin include soy milk, acorn squash, pistachio nuts, fortified foods, green peas, and watermelon. Thiamin supplements and the thiamin used in food fortification are usually in the form of *thiamin hydrochloride* or *thiamin nitrate*. Thiamin is not toxic in food or in the amounts found in most vitamin supplements.

Cooking can reduce thiamin in two ways. Thiamin is destroyed by heat. Also, thiamin is easily leached out of food by water, as seen in Figure 1-4. To minimize the loss of thiamin and other water-soluble vitamins during cooking, food can be steamed or made into stews and soups.

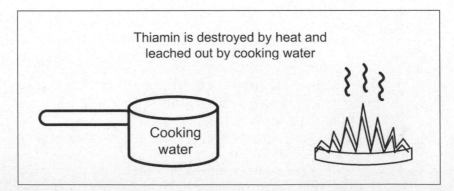

Figure 1-4   Thiamin can be lost in cooking.

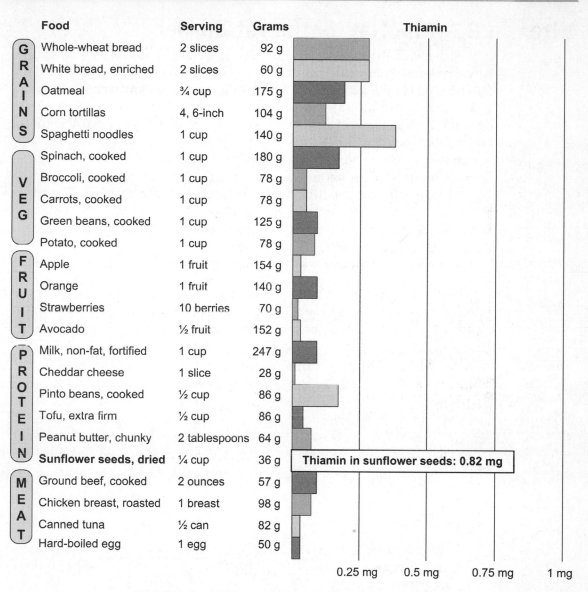

Food	Serving	Grams	Thiamin
GRAINS Whole-wheat bread	2 slices	92 g	
White bread, enriched	2 slices	60 g	
Oatmeal	¾ cup	175 g	
Corn tortillas	4, 6-inch	104 g	
Spaghetti noodles	1 cup	140 g	
VEG Spinach, cooked	1 cup	180 g	
Broccoli, cooked	1 cup	78 g	
Carrots, cooked	1 cup	78 g	
Green beans, cooked	1 cup	125 g	
Potato, cooked	1 cup	78 g	
FRUIT Apple	1 fruit	154 g	
Orange	1 fruit	140 g	
Strawberries	10 berries	70 g	
Avocado	½ fruit	152 g	
PROTEIN Milk, non-fat, fortified	1 cup	247 g	
Cheddar cheese	1 slice	28 g	
Pinto beans, cooked	½ cup	86 g	
Tofu, extra firm	½ cup	86 g	
Peanut butter, chunky	2 tablespoons	64 g	
Sunflower seeds, dried	¼ cup	36 g	Thiamin in sunflower seeds: 0.82 mg
MEAT Ground beef, cooked	2 ounces	57 g	
Chicken breast, roasted	1 breast	98 g	
Canned tuna	½ can	82 g	
Hard-boiled egg	1 egg	50 g	

0.25 mg   0.5 mg   0.75 mg   1 mg

Graph 1-1   Thiamin amounts in some common foods.

Vitamin B₂—Riboflavin, the Fat Burner

Riboflavin was discovered as a growth factor in the early nineteenth century. Riboflavin has a greenish-yellow color, which led to one of its early names, *vitamin G*. In 1935, riboflavin was first synthesized in the lab, and by 1938, its structure was determined. When it is taken in excess of needs, riboflavin is responsible for the bright yellow color of urine.

Riboflavin does its primary work as part of a coenzyme named *Flavin Adenine Dinucleotide* (FAD). Coenzymes derived from riboflavin are called *flavins*. Flavins are needed for the metabolism of carbohydrates, fats, and proteins. One of the ways that the thyroid gland controls metabolism is by regulating riboflavin's coenzyme activity. FAD is needed to prepare fatty acids for energy production in the mitochondria of all cells, as seen in Figure 1-5.

Flavins derived from riboflavin play a vital role in the metabolism and elimination of toxins, drugs, carcinogens, and steroid hormones. One flavin, the coenzyme FAD is also needed in the reduction and oxidation cycle of *glutathione,* as shown in Figure 1-6. This glutathione reduction and oxidation cycle has a major role in protecting us from free radicals such as hydrogen peroxide. When glutathione protects us from free radical oxidation, it becomes oxidized. FAD is needed to reduce (recharge) the glutathione and return the glutathione to its protective state.

Riboflavin deficiency is associated with the increased oxidative stress that can be caused by free radicals. A deficiency of riboflavin will reduce the efficiency of glutathione, an important antioxidant. In fact, measurement of glutathione reductase activity in red blood cells is used to assess the nutritional deficiency of riboflavin.

Another coenzyme made from riboflavin is *Flavin MonoNucleotide* (FMN). FMN is needed for the activation of *pyridoxine* (vitamin B₆). This is one of the

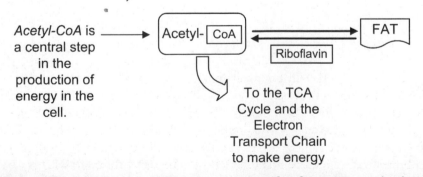

Figure 1-5 Riboflavin as FAD helps to prepare fats for energy production.

The free radical *hydrogen peroxide* is neutralized to water by *glutathione*. The *glutathione* is recharged by a *riboflavin-dependent coenzyme FAD*.

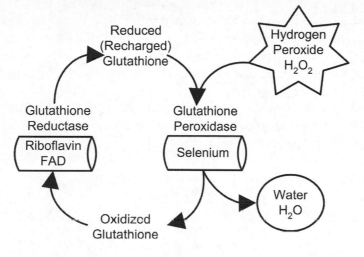

Figure 1-6   Riboflavin's role in antioxidant protection.

reasons that it is best to take all of the B vitamins together. A deficiency of riboflavin prevents the activation of pyridoxine, so taking pyridoxine by itself is not a good idea. Riboflavin in the forms of FMN and FAD coenzymes is also required by the mitochondrial electron transport chain for energy metabolism, as seen in Figure 1-7. Riboflavin is also needed by the *tricarboxylic acid* (TCA) cycle during energy production. This cycle has also been known as the *Krebs cycle*.

Riboflavin Coenzyme forms are:
Flavin Adenine Dinucleotide (FAD)
Flavin MonoNucleotide (FMN)

Riboflavin's recommended daily allowance (RDA) is primarily based upon its role as a coenzyme. Riboflavin needs will be met by most normal diets if enough food is eaten to meet energy requirements. Riboflavin is absorbed in the small intestines and released in the urine. People generally get enough riboflavin in the United States and Canada to meet the RDAs. The RDA for men is 1.3 mg per day and for women, 1.1 mg. Riboflavin is not toxic.

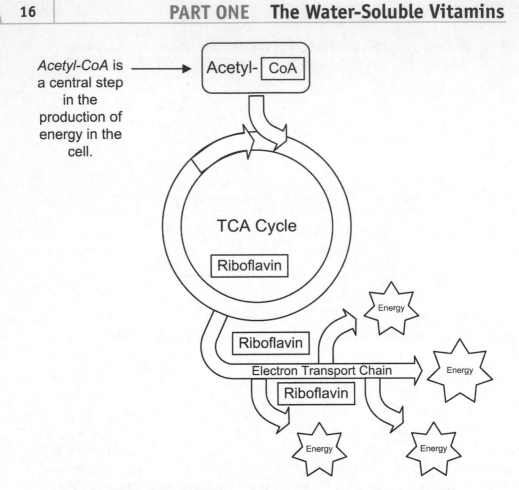

Figure 1-7   Riboflavin is needed for energy production in the cell.

No specific disease is caused by riboflavin deficiency. However, riboflavin deficiency can cause inflammation of the membranes of the eyes, the mouth, the skin, and the gastrointestinal tract. This condition is called *ariboflavinosis* (the prefix "a-" means *without* and the suffix "-osis" means *disease*). Riboflavin deficiency also can cause sensitivity to light. Cracks on the side of the mouth are another possible sign of riboflavin deficiency.

The best sources for riboflavin are whole grains and green leafy vegetables; please refer to Graph 1-2. Spinach, broccoli, chard, and asparagus are all rich sources of riboflavin. Almonds and soybeans are good sources. Dairy products have large amounts of riboflavin. Nutritional yeast is high in riboflavin and many other nutrients.

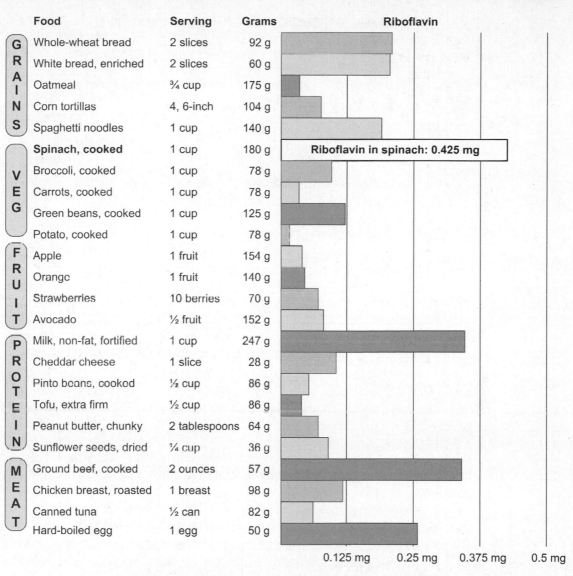

Food	Serving	Grams	Riboflavin
GRAINS			
Whole-wheat bread	2 slices	92 g	
White bread, enriched	2 slices	60 g	
Oatmeal	¾ cup	175 g	
Corn tortillas	4, 6-inch	104 g	
Spaghetti noodles	1 cup	140 g	
VEG			
Spinach, cooked	1 cup	180 g	Riboflavin in spinach: 0.425 mg
Broccoli, cooked	1 cup	78 g	
Carrots, cooked	1 cup	78 g	
Green beans, cooked	1 cup	125 g	
Potato, cooked	1 cup	78 g	
FRUIT			
Apple	1 fruit	154 g	
Orange	1 fruit	140 g	
Strawberries	10 berries	70 g	
Avocado	½ fruit	152 g	
PROTEIN			
Milk, non-fat, fortified	1 cup	247 g	
Cheddar cheese	1 slice	28 g	
Pinto beans, cooked	½ cup	86 g	
Tofu, extra firm	½ cup	86 g	
Peanut butter, chunky	2 tablespoons	64 g	
Sunflower seeds, dried	¼ cup	36 g	
MEAT			
Ground beef, cooked	2 ounces	57 g	
Chicken breast, roasted	1 breast	98 g	
Canned tuna	½ can	82 g	
Hard-boiled egg	1 egg	50 g	

0.125 mg   0.25 mg   0.375 mg   0.5 mg

Graph 1-2   Riboflavin amounts in some common foods.

Riboflavin is found in supplements in the form of riboflavin and in the form of *riboflavin monophosphate*. In supplements, riboflavin is most commonly found in vitamin B-complex preparations and in multivitamins.

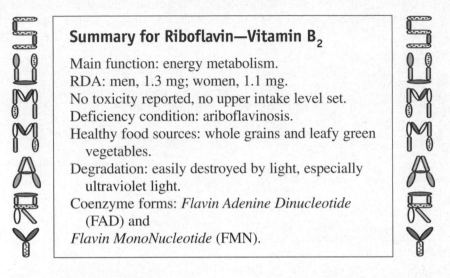

Summary for Riboflavin—Vitamin B$_2$

Main function: energy metabolism.
RDA: men, 1.3 mg; women, 1.1 mg.
No toxicity reported, no upper intake level set.
Deficiency condition: ariboflavinosis.
Healthy food sources: whole grains and leafy green vegetables.
Degradation: easily destroyed by light, especially ultraviolet light.
Coenzyme forms: *Flavin Adenine Dinucleotide* (FAD) and
Flavin MonoNucleotide (FMN).

Synthetic riboflavin used in supplements and fortified foods and drinks is very likely to have been produced using genetically modified *Bacillus subtilis*. These bacteria have been altered to increase the bacteria's production of riboflavin. The genes of the bacteria may also have been altered by the addition of antibiotic resistance to ampicillin. This is one possible difference between synthetic and naturally-occurring riboflavin.

Heat does not normally degrade riboflavin. However, ultraviolet light and other forms of irradiation including visible light destroy riboflavin, as shown in Figure 1-8.

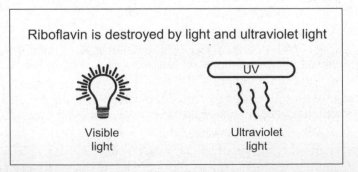

Figure 1-8   Riboflavin is destroyed by light and UV light.

Vitamin B$_3$—Niacin, Feel the Burn

The disease *pellagra* has been known since the introduction of corn to Europe in the 1700s. The connection between pellagra and niacin was confirmed in 1937 by an American scientist who was searching for the cause of pellagra.

Vitamin B$_3$ can be found in two different forms, *niacin* and *niacinamide*. Niacin is known chemically as *nicotinic acid* and can cause skin flushing if taken in larger doses. Niacinamide is the form of niacin commonly found in the blood and is known chemically as *nicotinamide*. Nicotinic acid can easily be converted into nicotinamide in the body. By the way, there is no chemical relationship to the nicotine in tobacco.

> Niacin   = Nicotinic Acid
> Niacinamide = Nicotinamide

Niacin is used in two coenzyme forms, *Nicotinamide Adenine Dinucleotide* (NAD) and *Nicotinamide Adenine Dinucleotide Phosphate* (NADP). Hundreds of enzymes require the niacin coenzymes, NAD and NADP. These enzymes are mainly used to accept or donate electrons to make energy or build molecules. NAD often functions in reactions involving the release of energy from carbohydrates, fats, proteins, and alcohol. Please note that NAD is used in many reactions involving energy production, as seen in Figure 1-9. NADP functions more often in biosynthesis, such as in the synthesis of fatty acids and cholesterol.

> Niacin Coenzyme forms are:
> *Nicotinamide Adenine Dinucleotide* (NAD)
> *Nicotinamide Adenine Dinucleotide Phosphate* (NADP)

Niacin can be made in the body from the essential amino acid *tryptophan*. When tryptophan is consumed in excess of protein needs, one sixtieth of the excess tryptophan can be converted to niacin. The *niacin equivalent* (NE) of food includes the niacin present plus the amount of niacin that can be made from tryptophan. In some foods, the tryptophan content is more important than the niacin content in providing niacin equivalents.

When a severe deficiency of niacin occurs, the deficiency disease is called *pellagra*. Pellagra is characterized by the four Ds: diarrhea, dermatitis, dementia, and death. Pellagra killed thousands of people in the South in the early twentieth

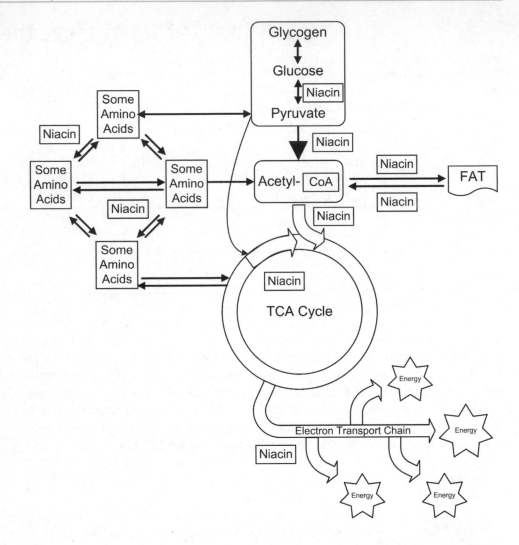

Figure 1-9   Niacin assists energy production from protein, fats, and carbohydrates.

century who were eating a diet consisting of mostly corn. Most of the niacin in corn is unavailable for easy assimilation, and corn is also low in tryptophan. In Mexico, pellagra is rare even in people on low-protein corn diets (tortillas, for example) because the corn is soaked in lime before the tortillas are cooked. This process releases the niacin for absorption.

Niacin is not toxic when obtained from food. The niacinamide (nicotinamide) form of niacin has not been found to be bothersome when taken in amounts less than 2000 mg per day. Decreased insulin sensitivity and liver toxicity can result from daily amounts over 2000 mg. However, niacin in the form of nicotinic acid can cause "*niacin flush*" in amounts as low as 35 mg. Nicotinic acid can dilate the capillaries, causing a brief tingling and flushing of the skin.

Summary for Niacin—Vitamin B$_3$

Main function: energy metabolism.
RDA: men, 16 mg; women, 14 mg (niacin equivalent).
No toxicity has been reported from food; flushing has
    been reported above 35 mg in the nicotinic acid form; no
    effects have been noted below 2000 mg of niacinamide.
Deficiency disease: pellagra.
Healthy food sources: whole grains and nuts.
Degradation: heat resistant.
Coenzyme forms: *Nicotinamide Adenine Dinucleotide*
    (NAD), *Nicotinamide Adenine Dinucleotide Phosphate*
    (NADP).

Nicotinic acid has been used to lower blood cholesterol in large doses of 3000 mg or more per day. This has resulted in a lowering of the bad LDL (low density lipoprotein) cholesterol and also raises the good HDL (high density lipoprotein) cholesterol. Liver damage and aggravation of diabetes are potential dangers of such high doses. People with a history of liver disease or abnormal liver function, diabetes, peptic ulcers, gout, cardiac arrhythmias, inflammatory bowel disease, migraine headaches, or alcoholism are more susceptible to the adverse effects of excessive nicotinic acid intake. Doses of 3000 mg or more per day are potentially dangerous therapies and must be supervised.

The average diet supplies an adequate amount of niacin. Niacin is not stored in the body. Cooked whole grains, legumes, and seeds are preferred sources of niacin; please refer to Graph 1-3. Enriched grains, mushrooms, leafy green vegetables, and nutritional yeast are other good sources. Pork, beef, chicken, fish, and dairy products are very high in niacin, but are also high in cholesterol.

Niacin is somewhat heat-resistant and is not depleted with normal cooking. Niacin may, however, be leached into cooking water. Niacin is also subject to losses during processing and storage, as seen in Figure 1-10.

The RDA for niacin is 16 mg for men and 14 mg for women. For the nicotinic acid form of niacin, the upper limit to avoid flushing is 35 mg per day. Niacin used in supplements and as food fortification is usually in the form of nicotinamide (niacinamide), which has no flushing effect. Nicotinic acid (niacin) is also available as a supplement, but should be kept to small doses to avoid flushing.

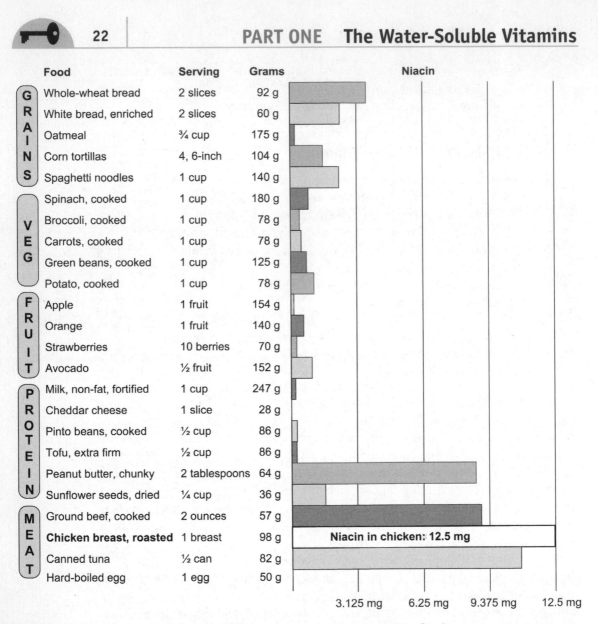

Food	Serving	Grams	Niacin
GRAINS			
Whole-wheat bread	2 slices	92 g	
White bread, enriched	2 slices	60 g	
Oatmeal	¾ cup	175 g	
Corn tortillas	4, 6-inch	104 g	
Spaghetti noodles	1 cup	140 g	
VEG			
Spinach, cooked	1 cup	180 g	
Broccoli, cooked	1 cup	78 g	
Carrots, cooked	1 cup	78 g	
Green beans, cooked	1 cup	125 g	
Potato, cooked	1 cup	78 g	
FRUIT			
Apple	1 fruit	154 g	
Orange	1 fruit	140 g	
Strawberries	10 berries	70 g	
Avocado	½ fruit	152 g	
PROTEIN			
Milk, non-fat, fortified	1 cup	247 g	
Cheddar cheese	1 slice	28 g	
Pinto beans, cooked	½ cup	86 g	
Tofu, extra firm	½ cup	86 g	
Peanut butter, chunky	2 tablespoons	64 g	
Sunflower seeds, dried	¼ cup	36 g	
MEAT			
Ground beef, cooked	2 ounces	57 g	
Chicken breast, roasted	1 breast	98 g	Niacin in chicken: 12.5 mg
Canned tuna	½ can	82 g	
Hard-boiled egg	1 egg	50 g	

3.125 mg   6.25 mg   9.375 mg   12.5 mg

Graph 1-3 Niacin amounts in some common foods.

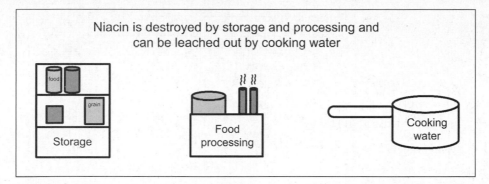

Figure 1-10   Niacin can leach into cooking water and can be lost during storage and processing.

Biotin—the Energy Catalyst

The name *biotin* is taken from the Greek word *bios*, which means life. Biotin was originally referred to as *vitamin H*. Biotin was discovered in late 1930s when animals developed skin problems when fed only egg whites. It took forty years of research to confirm biotin as a B vitamin because deficiency is so rare. Biotin is required by plants and animals. Biotin can only be synthesized by bacteria, algae, yeasts, molds, and a few plant species.

Biotin is used in four important enzymes, known as *carboxylases* (enzymes that donate carbon dioxide). *Acetyl-CoA carboxylase* is required for the synthesis of fatty acids. *Pyruvate carboxylase* is needed for the production of glucose from fats and proteins (called *gluconeogenesis*). *Methylcrotonyl-CoA carboxylase* is a catalyst in the metabolism of leucine, an essential amino acid. *Propionyl-CoA carboxylase* is needed for the metabolism of cholesterol, amino acids, and certain fatty acids.

Biotin Coenzyme forms are:
Acetyl-CoA carboxylase helps make fatty acids
Pyruvate carboxylase helps make blood sugar
   from fats and protein
Methylcrotonyl-CoA carboxylase helps
   metabolize the amino acid leucine
Propionyl-CoA carboxylase helps burn fats

Biotin is needed to make energy in the mitochondria of the cell as shown in Figure 1-11. Biotin may also play a role in the transcription and replication of DNA.

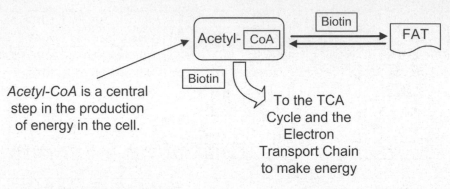

Figure 1-11   Biotin assists energy production in the cell.

Deficiency of biotin has been noted in prolonged intravenous feeding where biotin was omitted. The only other example of biotin deficiency is from long-term consumption of raw egg whites. Biotin can bind to *avadin*, a protein found in egg whites. Cooking inactivates this bond, so cooked egg whites do not bind biotin.

Deficiency can result from a genetic lack of *biotinidase*, an enzyme that releases biotin from small proteins. This lack of biotinidase is a rare hereditary disorder. Pregnant women may be at risk for borderline biotin deficiency. The developing fetus requires more biotin than is sometimes available. Also, large doses of pantothenic acid may compete with biotin for absorption in the intestines; this is because pantothenic acid and biotin have very similar structures.

Biotin is made by the bacteria living in healthy large intestines. The wall of the large intestine has a specialized process for the uptake of biotin. This may be one reason why biotin deficiency is so rare.

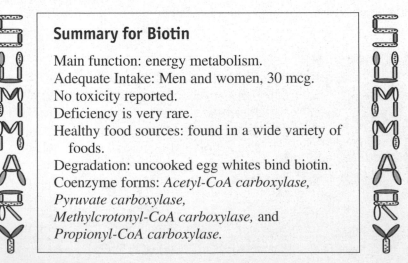

Summary for Biotin

Main function: energy metabolism.
Adequate Intake: Men and women, 30 mcg.
No toxicity reported.
Deficiency is very rare.
Healthy food sources: found in a wide variety of foods.
Degradation: uncooked egg whites bind biotin.
Coenzyme forms: *Acetyl-CoA carboxylase, Pyruvate carboxylase, Methylcrotonyl-CoA carboxylase,* and *Propionyl-CoA carboxylase.*

Adequate intake for biotin has been set at 30 mcg per day for healthy adults. Biotin is found in a wide variety of foods. Anticonvulsant medications may lower biotin levels. Additionally, antibiotics may lower the production of biotin because some antibiotics kill bacteria in the colon. Biotin is not known to be toxic.

Vitamin B₅—Pantothenic Acid, the Center of Energy

Pantothenic acid derives its name from the root word *pantos*, which means "everywhere." It has been found in every living cell including plant and animal tissues as well as in microorganisms. Pantothenic acid was first identified in 1933 when Roger Williams detected it as a growth factor for yeast. It is used in a wide variety of vital body processes.

Pantothenic acid forms a large part of the coenzyme A molecule. Coenzyme A is essential for the chemical reactions that generate energy from carbohydrates, fats, and proteins, as you can see in Figure 1-12.

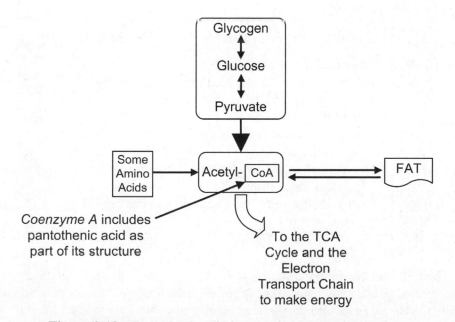

Figure 1-12   Pantothenic acid is central to energy production.

Pantothenic acid, in the form of coenzyme A, is needed for the synthesis of cholesterol and the synthesis of steroid hormones such as *melatonin*. Coenzyme A is also needed for the synthesis of *acetylcholine*, a neurotransmitter. *Heme*, a component of hemoglobin, cannot be synthesized without coenzyme A. In addition, the liver requires coenzyme A in order to metabolize a number of drugs and toxins.

> Pantothenic acid forms a large part of coenzyme A.

Pantothenic acid in the form of coenzyme A is indispensable for the synthesis of fats used in the *myelin sheaths* of nerve cells, and also synthesizes the *phospholipids* in cell membranes.

Pantothenic acid deficiency is very rare and seen only in cases of severe malnutrition. Pantothenic acid is found in many common foods and average diets are thought to have an adequate amount of it. Pantothenic acid is also made by the normal bacteria that live in the colon. Absorption of pantothenic acid from the colon has been demonstrated, but may not be available in meaningful amounts from colonic bacteria.

Summary for Pantothenic Acid—Vitamin B$_5$

Main function: Energy metabolism.

Adequate Intake Level: Men and women, 6 mg.

No toxicity or deficiency disease reported, no upper intake level set.

Healthy food sources: avocado, sunflower seeds, and sweet potatoes.

Degradation: easily destroyed by freezing, canning, and refining.

Coenzyme form: pantothenic acid forms a part of coenzyme A.

Healthy sources of pantothenic acid include whole grains, nuts and seeds, nutritional yeast, sweet potatoes, legumes, mushrooms, tomatoes, and broccoli; please refer to Graph 1-4. Enriched grains such as white flour are not

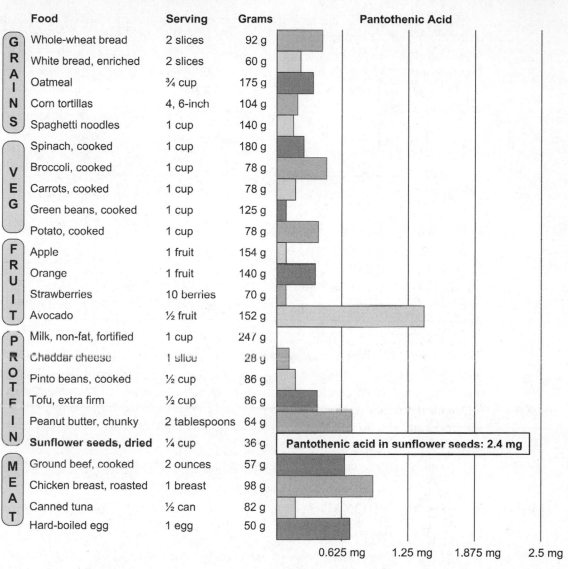

Food	Serving	Grams	Pantothenic Acid
GRAINS			
Whole-wheat bread	2 slices	92 g	
White bread, enriched	2 slices	60 g	
Oatmeal	¾ cup	175 g	
Corn tortillas	4, 6-inch	104 g	
Spaghetti noodles	1 cup	140 g	
VEG			
Spinach, cooked	1 cup	180 g	
Broccoli, cooked	1 cup	78 g	
Carrots, cooked	1 cup	78 g	
Green beans, cooked	1 cup	125 g	
Potato, cooked	1 cup	78 g	
FRUIT			
Apple	1 fruit	154 g	
Orange	1 fruit	140 g	
Strawberries	10 berries	70 g	
Avocado	½ fruit	152 g	
PROTEIN			
Milk, non-fat, fortified	1 cup	247 g	
Cheddar cheese	1 slice	28 g	
Pinto beans, cooked	½ cup	86 g	
Tofu, extra firm	½ cup	86 g	
Peanut butter, chunky	2 tablespoons	64 g	
Sunflower seeds, dried	¼ cup	36 g	Pantothenic acid in sunflower seeds: 2.4 mg
MEAT			
Ground beef, cooked	2 ounces	57 g	
Chicken breast, roasted	1 breast	98 g	
Canned tuna	½ can	82 g	
Hard-boiled egg	1 egg	50 g	

0.625 mg   1.25 mg   1.875 mg   2.5 mg

Graph 1-4   Pantothenic acid amounts in some common foods.

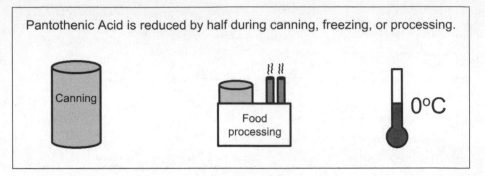

Figure 1-13　Pantothenic acid is reduced by processing, freezing, and canning.

enriched with pantothenic acid and about 43 percent of the pantothenic acid is lost in the milling process. Both freezing and canning decrease the pantothenic acid by approximately half, as seen in Figure 1-13. The adequate daily intake is 5 mg.

Pantothenic acid is not known to be toxic. It is easily eliminated in the urine. Oral contraceptives may increase the need for pantothenic acid. Supplements are usually in the form of *pantothenol*, a stable form of the vitamin. Supplements are also made from calcium and sodium *D-pantothenate*. The *panthene* form of pantothenic acid is a cholesterol-lowering drug used only under expert supervision.

Vitamin B$_6$—Pyridoxine, the Protein Burner

Vitamin B$_6$ was discovered in the 1930s. Vitamin B$_6$ occurs in several forms, all of which can be converted to the most active coenzyme form, *Pyridoxal Phosphate* (PLP). PLP has a vital role in catalyzing dozens of chemical reactions in the body.

Vitamin B$_6$ is unusual as a B vitamin in that it is so extensively stored in muscle tissue. Glucose is stored as *glycogen* in muscle tissue to provide energy on-site and when it is first needed. PLP is a coenzyme to the enzyme *glycogen phosphorylase*, which catalyzes the release of glucose from glycogen,

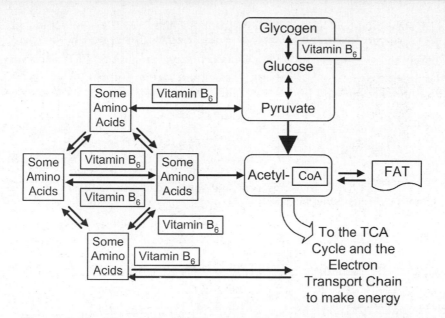

Figure 1-14   Vitamin B$_6$ (pyridoxine) assists energy production from protein and carbohydrates.

shown in the top of Figure 1-14. In another process, known as *gluconeogenesis* (gluco-neo-genesis means "sugar-new-making"), PLP is needed to convert amino acids to glucose.

Vitamin B$_6$ Coenzyme form is:
Pyridoxal Phosphate (PLP)

Many important neurotransmitters are synthesized using PLP-dependent enzymes. *Serotonin* is synthesized from tryptophan in the brain with the help of PLP. Other neurotransmitters that are synthesized using PLP-dependent enzymes are *dopamine, gamma-aminobutyric acid* (GABA), and *norepinephrine.*

In the discussion of niacin, it was noted that tryptophan can be converted to niacin. The vitamin B$_6$ coenzyme PLP is needed to convert tryptophan to niacin. This is another example of how all of the B vitamins work together.

PLP can help transfer amino groups from one amino acid to another in a process called *transamination* (moving amino acids). In this way PLP can help make non-essential amino acids in the body. Along with other nutrients, PLP also functions in the synthesis of heme, a part of hemoglobin. Some forms of vitamin B$_6$ help hemoglobin pick up and release oxygen.

> PLP from vitamin B$_6$ serves as a coenzyme in the synthesis of nucleic acids such as DNA (deoxyribonucleic acid) and RNA, (ribonucleic acid).

Vitamin B$_6$ is depleted by alcohol drinking. Alcohol is broken down to *acetaldehyde* in the body. Acetaldehyde breaks the PLP coenzymes loose from their enzymes and the PLP is lost. Some drugs, such as *isonicotinic acid hydrazide* (INH) also deplete the body of vitamin B$_6$. INH is a drug used to treat tuberculosis and supplemental vitamin B$_6$ must be given during treatment.

Elevated *homocysteine* in the blood is an indicator that there is increased risk of cardiovascular disease. Vitamin B$_6$ is needed to remove homocysteine from the blood by converting homocysteine to cysteine, as seen in Figure 1-15. In another process, folate and vitamin B$_{12}$ remove homocysteine from the blood by converting homocysteine to methionine.

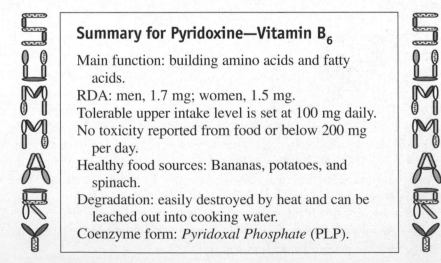

Summary for Pyridoxine—Vitamin B$_6$

Main function: building amino acids and fatty acids.

RDA: men, 1.7 mg; women, 1.5 mg.

Tolerable upper intake level is set at 100 mg daily.

No toxicity reported from food or below 200 mg per day.

Healthy food sources: Bananas, potatoes, and spinach.

Degradation: easily destroyed by heat and can be leached out into cooking water.

Coenzyme form: *Pyridoxal Phosphate* (PLP).

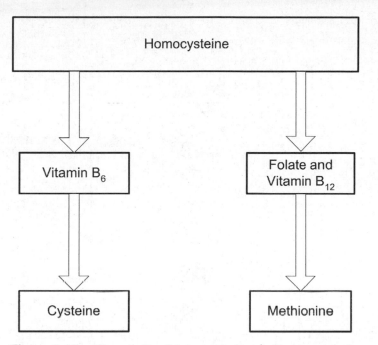

Figure 1-15   Vitamin B_6, folate, and vitamin B_{12} convert the undesirable homocysteine to the useful amino acids cysteine and methionine.

Adults need about 1.3 mg to 1.7 mg of vitamin B_6 to meet the RDAs. Severe deficiency is uncommon because vitamin B_6 is found in many foods and fortified in refined grains; please refer to Graph 1-5. Alcoholics are at risk of deficiency. Increased protein intake increases the need for vitamin B_6. Dietary intake of vitamin B_6 in the United States averages about 2 mg/day for men and 1.5 mg/day for women, thus meeting the RDAs. Bananas, fortified cereal, spinach, chicken, salmon, and potatoes are high in vitamin B_6. Vitamin B_6 is easily destroyed by heat, as shown in Figure 1-16.

Vitamin B_6 is not toxic when supplied by food in the diet. Supplemental forms are usually in the form of *pyridoxine hydrochloride*. The upper limit for adults set by the Institute of Medicine is 100 mg; 100 mg per day is certainly a safe limit, well below the level that might bring on neurological problems. Vitamin B_6 deficiency causes depression and confusion, and, in extreme deficiency, brain wave abnormalities and convulsions.

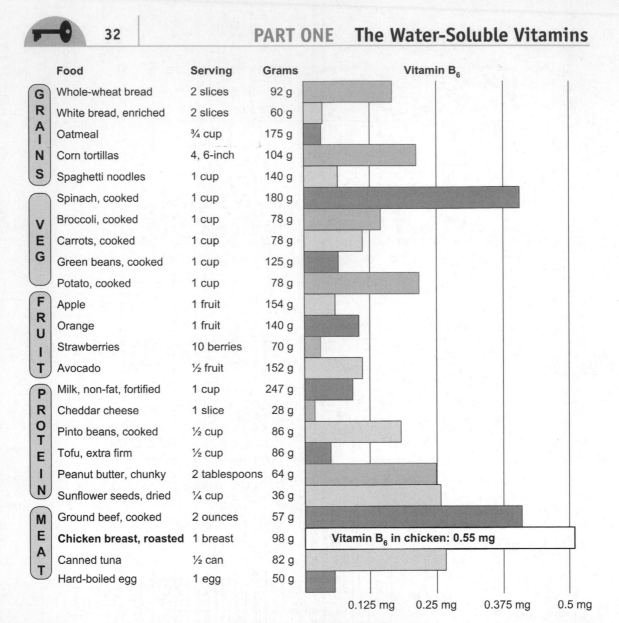

Food	Serving	Grams	Vitamin B$_6$
GRAINS			
Whole-wheat bread	2 slices	92 g	
White bread, enriched	2 slices	60 g	
Oatmeal	¾ cup	175 g	
Corn tortillas	4, 6-inch	104 g	
Spaghetti noodles	1 cup	140 g	
VEG			
Spinach, cooked	1 cup	180 g	
Broccoli, cooked	1 cup	78 g	
Carrots, cooked	1 cup	78 g	
Green beans, cooked	1 cup	125 g	
Potato, cooked	1 cup	78 g	
FRUIT			
Apple	1 fruit	154 g	
Orange	1 fruit	140 g	
Strawberries	10 berries	70 g	
Avocado	½ fruit	152 g	
PROTEIN			
Milk, non-fat, fortified	1 cup	247 g	
Cheddar cheese	1 slice	28 g	
Pinto beans, cooked	½ cup	86 g	
Tofu, extra firm	½ cup	86 g	
Peanut butter, chunky	2 tablespoons	64 g	
Sunflower seeds, dried	¼ cup	36 g	
MEAT			
Ground beef, cooked	2 ounces	57 g	
Chicken breast, roasted	1 breast	98 g	**Vitamin B$_6$ in chicken: 0.55 mg**
Canned tuna	½ can	82 g	
Hard-boiled egg	1 egg	50 g	

0.125 mg   0.25 mg   0.375 mg   0.5 mg

Graph 1-5   Vitamin B$_6$ amounts in some common foods.

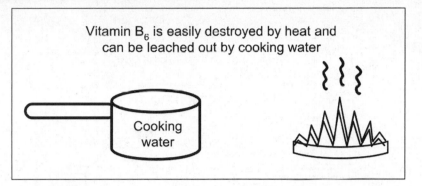

Figure 1-16   Vitamin B$_6$ is easily destroyed by heat and lost to cooking water.

Folate—the DNA Creator

In 1945 folic acid was isolated from spinach. The name *folate* is derived from the word *foliage*. The terms *folate* and *folic acid* are both used for this water-soluble B vitamin. Folic acid is the form normally used in vitamin supplements and in fortifying food. Folic acid is readily converted to folate in the body. Folic acid is rarely found in food or in the human body. On the other hand, folates are found in food in many naturally-occurring forms. Folates are the metabolically active forms in the human body.

FOLATE ENZYMES

The primary coenzyme form of folate is *TetraHydroFolate* (THF). THF is needed to transfer *one-carbon units*. These one-carbon units contain a single carbon atom, which may be added to a compound being biosynthesized. Folate coenzymes act as acceptors and donors of one-carbon units in a variety of reactions needed in the metabolism of amino acids (shown in Figure 1-17) and also nucleic acids.

Folate Coenzyme form is:
TetraHydroFolate (THF)

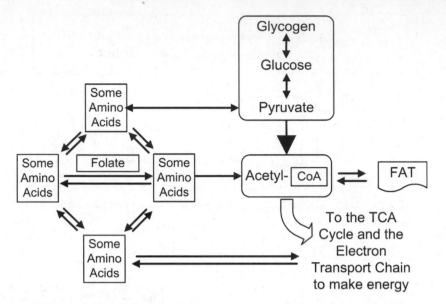

Figure 1-17   Folate is needed in the metabolism of amino acids.

One of the main roles of folate in the body is assisting in the metabolism of nucleic acids (DNA and RNA). Synthesis of DNA is dependent on folate co-enzymes. Synthesis of DNA is especially important in rapidly-growing cells, such as red blood cells.

FOLATE AND HOMOCYSTEINE

Folate, along with vitamin B_{12}, is needed in the synthesis of methionine. Methionine is needed to synthesize *S-adenosylmethionine* (SAMe), as seen in Figure 1-18. SAMe is used as a methyl donor at many sites within both DNA and RNA. A methyl donor is any substance that can transfer a methyl group (CH_3) to another substance. These methyl groups can protect DNA against the changes that might lead to cancer. This synthesis of methionine from homocysteine is also important to prevent a buildup of homocysteine in the blood.

Deficiency of folates in the blood results in increased homocysteine levels. These increased homocysteine levels increase the risk of fatal heart disease. Slightly elevated levels of homocysteine in the blood have been associated with thickening of arterial walls, arterial wall deterioration, and blood clot forma-

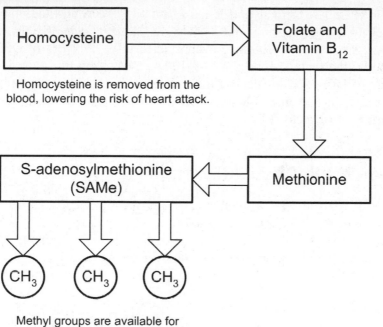

Figure 1-18 Folate and vitamin B$_{12}$ lower homocysteine levels and lead to the formation of SAMe, which donates methyl groups.

tion. Vitamin B$_6$ and vitamin B$_{12}$ are also needed to lower homocysteine levels in the blood.

> Elevated homocysteine levels in the blood have been associated with thickening of arterial walls, arterial wall deterioration, and blood clot formation.

BIOAVAILABILITY OF FOLATE AND FOLIC ACID

Synthetic folic acid commonly found in fortified food and in supplements has a higher bioavailability than the naturally occurring folates. It is unusual for a synthetic vitamin to be more potent that a naturally-occurring vitamin. For folates, 100 mcg (micrograms) is also 100 mcg of *Dietary Folate Equivalent* (DFE). However, for folic acid, 100 mcg is considered to be 170 mcg of DFE, if eaten with food. When eaten between meals, 100 mcg of folic acid is considered to be 200 mcg of DFE. So, to calculate the total amount of folates, one must add the DFE from folates to the DFE from folic acid supplementation (normally using the multiplication factor 1.7).

ASSIMILATION OF FOLATE

Folate in food is normally bound to a string of amino acids; this is called the *polyglutamate* form. For intestinal absorption, it is necessary for folate to be bound to only one amino acid; this is called the *monoglutamate* form and is shown in Figure 1-19. Digestive enzymes on the surface of intestinal cells hydrolyze

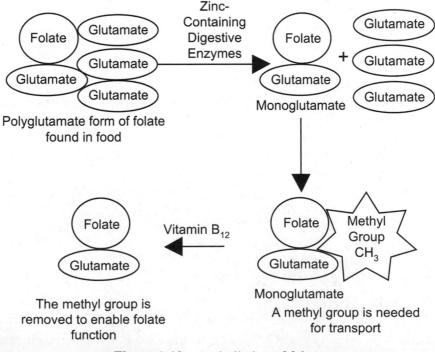

Figure 1-19   Assimilation of folate.

(split) the polyglutamate into one monoglutamate and several glutamates. A methyl group attaches to this monoglutamate to enable the transport of folate to the blood, liver, and other cells. This methyl group is needed for transport, but the methyl group needs to be removed for folate coenzymes to function. Vitamin B_{12} is part of the enzyme system that removes this methyl group from folate, thus activating the folate.

RECYCLING OF FOLATE

When there is excess folate in the blood, the liver puts much of the extra folate into bile. This bile goes back into the intestines, where the folate may be reabsorbed. This reabsorption of folate is dependent on healthy intestines.

This reabsorption of folate by the intestines is vital. If the intestines are not healthy enough to absorb folate, then this limits the ability of the body to make rapidly-growing cells. The intestinal walls are made up of rapidly-growing cells. In a vicious cycle, the lack of folate can prevent uptake of folate and other nutrients. Also, alcohol drinking can decrease the absorption of folate.

NEURAL TUBE DEFECTS

Lack of folate in the diet can cause *neural tube defects* in an embryo of a pregnant woman. These can cause devastating and sometimes fatal birth defects. Five servings of fruits and vegetables daily will provide enough folate for normal embryonic development. However, many American women do not eat this quantity of fruits and vegetables. To be assured of adequate folate to prevent neural tube defects, women are advised to take 400 mcg (micrograms) of folic acid daily either as a dietary supplement or in fortified food. This folic acid supplementation needs to be taken from one month before conception until the end of the first trimester of pregnancy. To protect against neural tube defects, the U.S. Food and Drug Administration has mandated that folic acid be added to enriched grain products. This enrichment of grain with folate is estimated to deliver an average of 100 mcg of folic acid daily.

> Folate is needed to prevent birth defects called *neural tube defects*.

FOLATE AND CANCER

Folate may also play a role in cancer prevention. Cancer is thought to arise when DNA is damaged faster than it can be repaired. Folate assists with synthesis and repair of DNA. Folate also assists with *methylation* by increasing the SAMe in the cell.

Cancer is thought to arise from excess expression of certain genes within DNA. Methylation quiets RNA replication and can help the cells regulate excess growth. Folate may be helpful in preventing breast cancer in women who drink alcohol. Sufficient folate intake has also been associated with lower colorectal cancer risk.

FOLATE AND VITAMIN B$_{12}$

High levels of folic acid supplementation, especially above 1000 mcg daily, can mask symptoms of vitamin B$_{12}$ deficiency. The addition of folic acid to food is risky because some people may get enough folic acid to hide the terrible effects of vitamin B$_{12}$ deficiency on nerves. The addition of folic acid to grains is also controversial because it only helps prevent neural tube defects about half of the time.

FOLATE DEFICIENCY

Folate deficiency signs include anemia and deterioration of the gastrointestinal tract. The anemia results from abnormal blood cell division resulting in fewer and larger red blood cells. This type of anemia is called *megaloblastic* anemia, referring to the large immature red blood cells. *Neutrophils*, a type of white blood cell, also can be abnormal. Symptoms of folate deficiency typically take a few months to develop and may result in fatigue from lower oxygen-carrying capacity of red blood corpuscles.

Folate deficiency can be caused by inadequate absorption, inadequate consumption, or unusually high metabolic needs for this vitamin. Drugs that can interfere with folate include aspirin, anticancer drugs, antacids, and oral contraceptives. Older people tend to have higher homocysteine levels and are encouraged to meet or exceed the RDA for folate.

FOLATE IN FOOD

Leafy green vegetables are the best sources of folate. Legumes are also high in folate; please refer to Graph 1-6. Other sources include fruit and enriched grains.

Summary for Folate

Main function: synthesis of DNA and red blood
   cells.
RDA: adults, 400 mcg; pregnant women, 600 mcg.
Tolerable upper intake level: 1000 mcg.
No toxicity reported from food.
Deficiency disease: neural tube defects, anemia, and
   excess folates may mask vitamin B_{12} deficiency.
Healthy food sources: leafy green vegetables and
   legumes.
Degradation: easily destroyed by oxygen and heat.
Coenzyme form: *TetraHydroFolate* (THF).

The RDA for folate as expressed in DFE (Dietary Folate Equivalents) is 400 mcg
per day for adults. A tolerable upper intake level of 1,000 mcg per day was established by the Food and Nutrition Board to prevent the masking of vitamin B_{12} deficiency. Folate and folic acid are not toxic.

    Folate in food can be destroyed by heat, light, and air, as shown in Figure 1-20.
Among the various cooking techniques, microwave cooking results in the greatest
losses of folate.

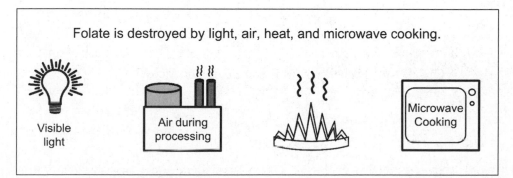

Folate is destroyed by light, air, heat, and microwave cooking.

Visible light

Air during processing

Microwave Cooking

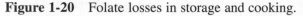

Figure 1-20   Folate losses in storage and cooking.

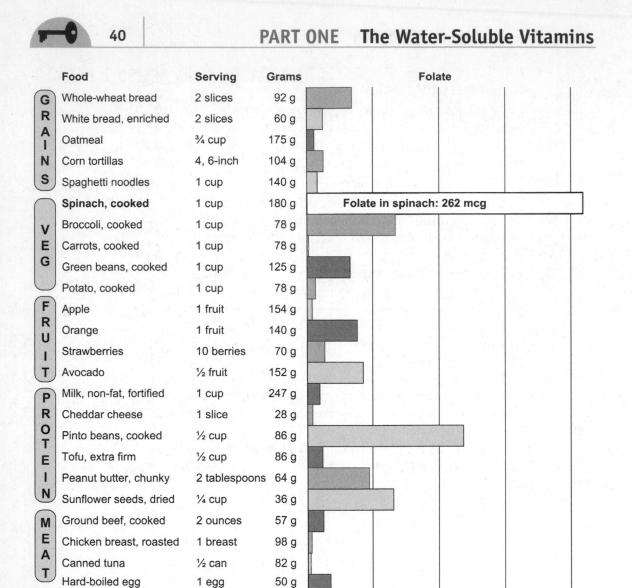

Food	Serving	Grams	Folate
GRAINS			
Whole-wheat bread	2 slices	92 g	
White bread, enriched	2 slices	60 g	
Oatmeal	¾ cup	175 g	
Corn tortillas	4, 6-inch	104 g	
Spaghetti noodles	1 cup	140 g	
VEG			
Spinach, cooked	1 cup	180 g	Folate in spinach: 262 mcg
Broccoli, cooked	1 cup	78 g	
Carrots, cooked	1 cup	78 g	
Green beans, cooked	1 cup	125 g	
Potato, cooked	1 cup	78 g	
FRUIT			
Apple	1 fruit	154 g	
Orange	1 fruit	140 g	
Strawberries	10 berries	70 g	
Avocado	½ fruit	152 g	
PROTEIN			
Milk, non-fat, fortified	1 cup	247 g	
Cheddar cheese	1 slice	28 g	
Pinto beans, cooked	½ cup	86 g	
Tofu, extra firm	½ cup	86 g	
Peanut butter, chunky	2 tablespoons	64 g	
Sunflower seeds, dried	¼ cup	36 g	
MEAT			
Ground beef, cooked	2 ounces	57 g	
Chicken breast, roasted	1 breast	98 g	
Canned tuna	½ can	82 g	
Hard-boiled egg	1 egg	50 g	

62.5 mcg   125 mcg   187 mcg   250 mcg

Graph 1-6   Folate amounts in some common foods.

Vitamin B$_{12}$—Cobalamin, the Blood Maker

Vitamin B$_{12}$ is unusual for a vitamin in that it contains cobalt. The name, *cobal-amin*, derives from "cobal" from the word *cobalt* plus "amin" from the word *vitamin* (cobal-amin). During the 1930s, researchers were trying to identify an "antipernicious anemia factor." Vitamin B$_{12}$ was discovered simultaneously by two research teams, one in the United States and one in England. It was not until 1956 that the chemical structure of this complex vitamin was revealed.

Compounds having B$_{12}$ activity have cobalamin as part of their name. In the human body, the active coenzyme forms of vitamin B$_{12}$ are *methylcobalamin* and *deoxyadenosyl cobalamin*. Vitamin B$_{12}$ is the most complex of the vitamins. The crystals are a striking dark red in color.

> Cobalamin Coenzyme forms are:
> *Methylcobalamin*
> *Deoxyadenosyl cobalamin*

Vitamin B$_{12}$ supplements are usually in the form of *cyanocobalamin*, which is easily converted into the two active forms used in the body.

ROLES OF VITAMIN B$_{12}$

One function of vitamin B$_{12}$ is to convert homocysteine to methionine. One of the active forms of cobalamin, methylcobalamin, converts homocysteine using an enzyme that also requires folate. When folate gives up a methyl group, the methyl-cobalamin coenzyme becomes reactivated. This is important because, as mentioned, homocysteine is a risk factor for cardiovascular disease. The methionine that is converted can go on to become part of SAMe. SAMe is a methyl group donor that can protect DNA and may be important in cancer prevention.

The other active form of vitamin B$_{12}$, deoxyadenosyl cobalamin, has an important role in the production of energy from proteins and fats as you can see from Figure 1-21. Vitamin B$_{12}$ is essential for the maintenance of the nervous system and for the synthesis of molecules involved in fatty acid biosynthesis. Vitamin B$_{12}$ is needed to maintain the myelin sheath that surrounds nerve cells.

> Vitamin B$_{12}$ is needed to make red blood cells.

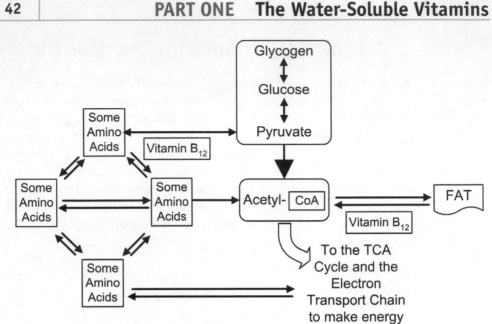

Figure 1-21   Vitamin B_{12} in energy metabolism.

Vitamin B_{12} is one of the nutrients required for the synthesis of hemoglobin, the oxygen-carrying pigment in blood. Vitamin B_{12} is needed for DNA synthesis in the rapidly dividing cells of the bone marrow. Lack of vitamin B_{12} or folate can lead to the production of large, immature, hemoglobin-poor red blood cells. This can result in *pernicious anemia*.

ASSIMILATION OF VITAMIN B_{12}

Vitamin B_{12} is normally bound to protein in food. Vitamin B_{12} is released from a protein by the digestive action of hydrochloric acid and pepsin as seen in Figure 1-22. The stomach then makes an intrinsic factor that binds to the vitamin B_{12}. In the intestines, the complex of intrinsic factor and vitamin B_{12} is slowly absorbed into the bloodstream. Once in the blood, transport of vitamin B_{12} is dependent on specific binding proteins.

Circulating vitamin B_{12} is recovered by the liver and returned to the intestines in bile. Most of this vitamin B_{12} is reabsorbed from the intestines. Because of this efficient recycling, most people do not develop a deficiency, even with minimal intake.

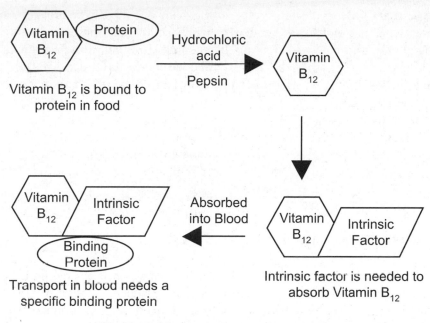

Figure 1-22   Assimilation of cobalamin.

DEFICIENCY OF VITAMIN B$_{12}$

Deficiency of vitamin B$_{12}$ is not usually from lack of intake, but rather from lack of absorption. Inadequate hydrochloric acid in the stomach will prevent vitamin B$_{12}$ from being released from dietary proteins so it cannot be utilized. Lack of intrinsic factor can also prevent absorption.

Pernicious anemia results from inadequate absorption of vitamin B$_{12}$, which can be caused by damaged stomach cells. It is most common in those over 60 years of age. The damaged stomach cells cannot make enough hydrochloric acid and/or enough intrinsic factor. Even a more than adequate dietary intake will not help because the vitamin B$_{12}$ is not absorbed through the intestines. With pernicious anemia, Vitamin B$_{12}$ must be injected or absorbed through membranes directly into the bloodstream via a nasal spray or absorbed from under the tongue.

It may take several years to develop deficiency symptoms as vitamin B$_{12}$ is efficiently recycled. Whether the deficiency is from poor absorption or from a diet low in animal products, onset of symptoms is typically slow.

Folate cannot be properly utilized if vitamin B$_{12}$ is low or absent. This is because vitamin B$_{12}$ is needed to convert methyl folate to its active form. If folate or vitamin B_{12} is absent, symptoms of folate anemia can be present. Folate anemia causes slowed DNA synthesis, which shows up first as defective red blood cells. Extra folate will clear up the anemia. However, if the anemia is due to lack of vitamin B_{12}, then the nerves will continue to suffer damage from lack of vitamin B_{12}. In this way folate can be dangerous because it can mask a vitamin B_{12} deficiency.

Deficiency of vitamin B_{12} first shows as a paralysis that begins in the extremities and moves inward. Correct identification of vitamin B_{12} deficiency can prevent permanent paralysis and nerve damage. Remember that there are no symptoms of anemia when folate levels are high and only the vitamin B_{12} levels are low.

VITAMIN B_{12} IN FOOD

Adults need only 2.4 micrograms (2.4 millionths of a gram) of vitamin B_{12} each day. This tiny amount, still billions of molecules, provides enough for all of our needs. Vitamin B_{12} is found only in foods made from animal products, such as meat, dairy products, and eggs; please refer to Graph 1-7. It is also available from nutritionally fortified yeast, although it is not found in brewer's yeast. Vitamin B_{12} is also found in many fortified vegetarian foods, such as soy milk. Vitamin B_{12} is made by bacteria in the colon. More research is needed to determine if enough vitamin B_{12} can be absorbed from the colon to be of significant value.

Vitamin B_{12} is easily lost through leaching into water while cooking. One unusual thing about vitamin B_{12} is that microwave cooking inactivates it, as shown in Figure 1-23. Cooking in the oven or in soups and stews preserves the most vitamin B_{12}.

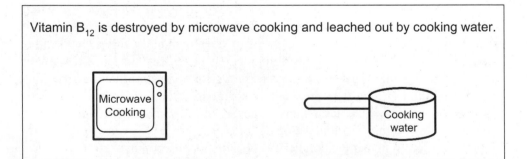

Figure 1-23   Some cooking methods reduce vitamin B_{12} content.

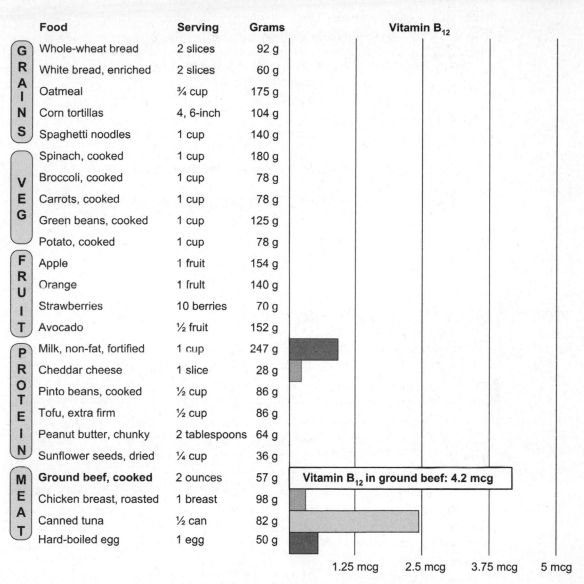

	Food	Serving	Grams	Vitamin B₁₂
GRAINS	Whole-wheat bread	2 slices	92 g	
	White bread, enriched	2 slices	60 g	
	Oatmeal	¾ cup	175 g	
	Corn tortillas	4, 6-inch	104 g	
	Spaghetti noodles	1 cup	140 g	
VEG	Spinach, cooked	1 cup	180 g	
	Broccoli, cooked	1 cup	78 g	
	Carrots, cooked	1 cup	78 g	
	Green beans, cooked	1 cup	125 g	
	Potato, cooked	1 cup	78 g	
FRUIT	Apple	1 fruit	154 g	
	Orange	1 fruit	140 g	
	Strawberries	10 berries	70 g	
	Avocado	½ fruit	152 g	
PROTEIN	Milk, non-fat, fortified	1 cup	247 g	
	Cheddar cheese	1 slice	28 g	
	Pinto beans, cooked	½ cup	86 g	
	Tofu, extra firm	½ cup	86 g	
	Peanut butter, chunky	2 tablespoons	64 g	
	Sunflower seeds, dried	¼ cup	36 g	
MEAT	**Ground beef, cooked**	2 ounces	57 g	Vitamin B₁₂ in ground beef: 4.2 mcg
	Chicken breast, roasted	1 breast	98 g	
	Canned tuna	½ can	82 g	
	Hard-boiled egg	1 egg	50 g	

1.25 mcg   2.5 mcg   3.75 mcg   5 mcg

Graph 1-7   Vitamin B$_{12}$ amounts in some common foods.

> **Summary for Cobalamin—Vitamin B$_{12}$**
>
> Main function: cell synthesis and red blood cells.
> RDA: men and women, 2.4 mcg.
> No toxicity reported from food or supplements.
> Deficiency disease: pernicious anemia.
> Healthy food sources: fortified cereals, nutritional yeast.
> Degradation: easily destroyed by microwave cooking.
> Coenzyme forms: *methylcobalamin* and *deoxyadenosyl cobalamin*.

How the B Vitamins Make Energy Production Possible

The main function of the B-vitamin complex is to assist in the production of energy in the body. They have many vital roles in energy production. Please study Figure 1-24. You have been seeing pieces of this diagram throughout this chapter. Figure 1-24 shows all of the B vitamins involved in energy production in one figure.

Many of the important roles of B vitamins take place in tiny organelles called mitochondria, which live deep inside the cells. Energy production in the mitochondria can be thought of in four steps. In each step, coenzymes made with B vitamins are essential. These coenzymes are discussed in the text and will be abbreviated below. The key to the enzyme abbreviations is in Figure 1-24.

STEP 1, BREAKING DOWN BLOOD SUGAR

The first step starts with glycogen, which, as you may recall, is the storage form of glucose, blood sugar. This first step breaks the glycogen down to glucose. This is done with the help of vitamin B$_6$ (PLP). Next, glucose is changed into pyruvate. This is done with the assistance of niacin (NAD). Pyruvate can also be made from certain amino acids with the help of vitamin B$_6$ (PLP).

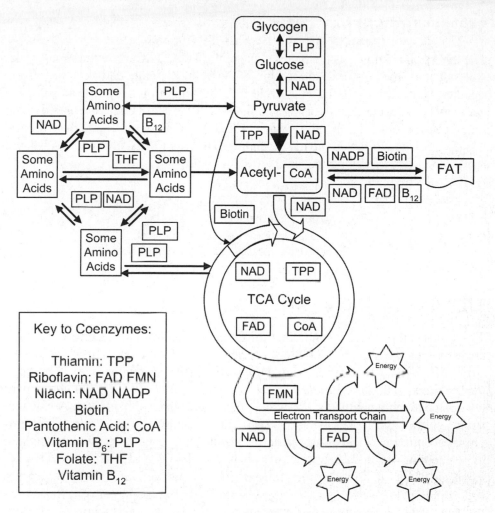

Figure 1-24   Energy production and the B vitamins.

STEP 2, THE CENTRAL COENZYME

The second step in energy production produces acetyl-coenzyme A. Pyruvate is made into acetyl-coenzyme A with the help of enzymes made from niacin (NAD) and thiamin (TPP). Acetyl-coenzyme A can also be made from some amino acids with a little help from folate (THF), niacin (NAD), and vitamin B_6 (PLP). Vitamin B_{12} may also be involved in amino acid transformation to ready the amino acids for this step. Fats may be converted to acetyl-coenzyme A with the help of niacin (NADP and NAD), riboflavin (FAD), vitamin B_{12}, and biotin.

STEP 3, THE ENERGY CYCLE

The third step is known as the Krebs cycle or the TCA (TriCarboxylic Acid) cycle. Acetyl-coenzyme A is used to feed the TCA cycle. Pyruvate can skip the second step and move directly to the TCA cycle with the help of biotin. Some amino acids get to the TCA cycle directly using vitamin B_6 (PLP). The TCA cycle uses oxygen to produce energy aerobically. Niacin (NAD), thiamin (TPP), and riboflavin (FAD) are needed to keep the cycle running. Thiamin (THF) and vitamin B_{12} are used to boost other compounds into the TCA cycle.

> Adenosine triphosphate (ATP) is the fully charged energy battery of the cell.

STEP 4, PUMPING UP ATP

The fourth step is called the electron transport chain. The electron transport chain pumps up the energy battery of the body, which is known as *adenosine triphosphate* (ATP). *Adenosine monophosphate* (*mono* means one) is pumped up to *adenosine diphosphate* (*di* means two), which is then pumped up to adenosine triphosphate (ATP), which has three (*tri*) phosphates. Adenosine triphosphate (ATP) is the fully charged energy battery of the cells. The high-energy bonds in ATP that are made in this step are used to produce energy in every cell in the body. Riboflavin (FMN and FAD) and niacin (NAD) play important roles in the electron transport chain, as you can see by reviewing Figure 1-24.

Many other nutrients play important roles in energy production in the cell. However, the B vitamins are the stars in energy production. The B vitamins work as coenzymes to do their indispensable work.

"Wanna B" Vitamins That Might Not Be Vitamins

Many substances are referred to as vitamins. Many of these substances are clearly not vitamins, as they are not essential for life. Some, such as choline, may one day

be called vitamins. While not essential for life, some of these substances may enhance health and disease resistance. Many of these "wanna B's" can be found in nutritional supplements.

CHOLINE

Choline is an essential nutrient that may not, strictly speaking, be a vitamin. Choline is synthesized in the body, but enough is not always made to meet needs. In order to make choline, the body needs sufficient methionine, vitamin B_{12}, and folic acid. Consequently, choline is sometimes needed in the diet. Lecithin (phosphatidyl-choline) contains about 13 percent choline by weight.

Most of the choline in the body is found in phospholipids. All human cell membranes need phospholipids for structural integrity. Choline is also a precursor for acetylcholine. Acetylcholine is an important neurotransmitter involved in muscle control, and memory. Large doses of lecithin have improved memory, but have not been helpful with Alzheimer's disease.

Without choline to make lecithin, the liver cannot rid itself of fats and cholesterol. This can lead to a condition known as "fatty liver." Choline is needed for liver health and liver damage results from deficiency.

Some choline is oxidized in the body to a metabolite known as *betaine*. Betaine supplies methyl groups for various methylation reactions. One of these methylation reactions results in the conversion of homocysteine to methionine.

No RDA has been set for choline, but the adequate daily intake (AI) has been set at 550 mg for men and 425 mg for women. Average dietary intake is thought to be adequate. Healthy food sources include Brussels sprouts, broccoli, peanut butter, and salmon.

Choline supplements are available as choline chloride and choline bitartrate. Lecithin is a good source of choline and contains 13 percent choline. Commercially available lecithin products may only contain 3–12 percent of choline because they often contain impure lecithin. The tolerable upper intake level for choline is 3.5 grams per day for adults. Very high levels of choline can disturb the neurotransmitter balance in the brain.

L-CARNITINE

L-carnitine is an essential nutrient that is also synthesized in well-nourished bodies. L-carnitine is a derivative of the essential amino acid lysine. L-carnitine was named after meat (carnus) because it was first isolated from meat in 1905. Although it is not officially a vitamin, carnitine has been called vitamin B_T.

L-carnitine is important in energy metabolism. It assists activated fatty acids into the mitochondria for energy production. It also assists in transporting metabolic debris out of the mitochondria.

L-carnitine synthesis in the body requires several nutrients to be successful. Needed for synthesis are methionine, lysine, iron, vitamin B_6, niacin, and vitamin C. One of the early symptoms of vitamin C deficiency is fatigue, which may be related to decreased synthesis of L-carnitine.

ALPHA-LIPOIC ACID

Alpha-lipoic acid is a naturally occurring compound that is synthesized in small amounts in the body. Alpha-lipoic acid is needed for several important mitochondrial enzyme complexes for energy production and the breakdown of amino acids. An antioxidant itself, alpha-lipoic acid is able to recharge other antioxidants such as vitamin C and glutathione. Lipoic acid, like thiamin and biotin, contains sulfur. Alpha-lipoic acid is approved for the treatment of diabetic neuropathy and is available by prescription in Germany.

BIOFLAVONOIDS

Bioflavonoids (or just *flavonoids*) have been referred to as vitamin P. Bioflavonoids are not vitamins and are not essential for life. Bioflavonoids are excellent antioxidants found in many plant foods. As potent antioxidants, they protect us from arterial damage and cancer formation. Parsley and elderberry have the highest content of bioflavonoids.

COENZYME Q$_{10}$

Coenzyme Q_{10} is not a vitamin, but plays many necessary biological roles in the body. It can be synthesized in the body and so it cannot be considered a vitamin. It is a powerful fat-soluble antioxidant found in virtually all cell membranes. Coenzyme Q_{10} is needed for mitochondrial energy production.

INOSITOL

Inositol can be made in the body from glucose, so it is not a vitamin. Inositol consumption from the average diet is about one gram daily. In the form of phosphatidylinositol, inositol makes up a small but important part of cell membranes. Inositol is widely found in cereals and legumes and is a component of dietary fiber.

OROTIC ACID

Orotic acid is sometimes known as vitamin B_{13} and is sometimes supplemented as calcium orotate. Orotic acid has not been proven to be essential for human nutrition so it is not officially a vitamin. Its functions are similar to vitamin B_{12} and it enhances the usefulness of vitamin B_{12} and folate. Many of the vitamin-like effects of orotic acid are undoubtedly due to its role in RNA and DNA synthesis. Orotic acid acts as one of the best mineral transporters. It is able to transport minerals such as calcium though the cell membrane and directly to the mitochondria.

PARA-AMINOBENZOIC ACID (PABA)

Para-Aminobenzoic Acid (PABA) is sometimes referred to as a B vitamin. It is not a vitamin and is not essential for nutrition. It is a non-protein amino acid. PABA is found in many sunscreen lotions.

VITAMIN B_{15}

Vitamin B_{15} has not been established as necessary for human nutrition and is not a vitamin. In the form of dimethylglycine it may enhance oxygen transport to the mitochondria.

VITAMIN B_{17}

Vitamin B_{17} (amygdalin) is not a vitamin. It is not necessary for human nutrition. It is found in apricot kernels and other nuts and seeds. It has been used for tumor therapy with much criticism and controversy.

VITAMIN F

Vitamin F has been one alternate name for the essential fatty acids (linoleic acid and linolenic acid). Since these fatty acids are not synthesized in the body and are required in the diet they may someday be considered vitamins. They are essential for life.

VITAMIN T

Vitamin T is not a vitamin, but a nutritional component of sesame seeds.

VITAMIN U

Vitamin U is not a vitamin, but an anti-ulcer factor in cabbage.

Quiz

Refer to the text in this chapter if necessary. A good score is at least 8 correct answers out of these 10 questions. The answers are listed in the back of this book.

1. The B vitamins are vital for:
   (a) Energy metabolism.
   (b) The transfer of amino groups from one amino acid to another.
   (c) The synthesis of neurotransmitters.
   (d) All of the above.

2. The best cooking method to preserve all of the B vitamins is:
   (a) Boiling.
   (b) Steaming.
   (c) Microwave cooking.
   (d) Deep frying.

3. Pyridoxine is:
   (a) Vitamin B_1.
   (b) Vitamin B_3.
   (c) Vitamin B_6.
   (d) Vitamin B_{12}.

4. Niacin in the form of nicotinic acid, when taken in large doses, can cause:
   (a) Lowered energy production.
   (b) Loss of weight.
   (c) Flushing of skin.
   (d) Higher total blood cholesterol.

5. The B vitamins that can be made by bacteria in the large intestines:

    (a) Biotin.

    (b) Vitamin B_{12}.

    (c) Pantothenic acid.

    (d) All of the above.

6. Transamination is:

    (a) A trancelike state resulting from overdoses of folic acid.

    (b) The synthesis of amino acids from parts of other amino acids.

    (c) The synthesis of neurotransmitters.

    (d) The animated state that occurs when all B vitamins are consumed.

7. The following two B vitamins are needed to keep red blood cells healthy:

    (a) Vitamin B_{12} and folic acid.

    (b) Vitamin B_6 and thiamin.

    (c) Vitamin B_{12} and riboflavin.

    (d) Vitamin B_6 and riboflavin.

8. The B vitamin that helps prevent neural tube defects during pregnancy is:

    (a) Vitamin B_6.

    (b) Pantothenic acid.

    (c) Biotin.

    (d) Folic acid.

9. Glycogen is:

    (a) Vitamin B_{13}.

    (b) The storage form of blood sugar.

    (c) The product of the electron transport chain.

    (d) Vitamin B_5.

10. The energy battery of the body, when fully charged is:

    (a) AMP, adenosine monophosphate.

    (b) ADP, adenosine diphosphate.

    (c) ATP, adenosine triphosphate.

    (d) All of the above.

CHAPTER 2

Vitamin C
The Citrus Antioxidant

The History of Vitamin C

Lack of vitamin C for long periods of time can result in *scurvy*. Scurvy was first noticed during long sea voyages. The cause of the disease was not initially known. In 1536, a French explorer had his crew stricken with scurvy while exploring the St. Lawrence River. He learned from local Indians that the tips of young arbor vitae evergreen needles cure scurvy. By 1617, John Woodall, a surgeon for the British East India Company, published a cure for scurvy—lemon juice.

James Lind was another British surgeon. He wrote *Treatise on the Scurvy* in 1753. James Lind gave some sailors two oranges and one lemon each day, while other sailors received cider, vinegar, or other possible scurvy cures. This may have been the first scientific nutrition experiment in the history of science. He proved that fresh citrus fruit prevented and cured scurvy. By 1795, limes were standard supplements on British ships and scurvy was no longer a problem. British seamen are called "limeys" to this day because of this custom. Captain James Cook sailed to the Hawaiian Islands using sauerkraut for his vitamin C and lost no men to scurvy.

DISCOVERY OF ASCORBIC ACID

Until the early twentieth century, the factor in these foods that prevented scurvy was an unknown *antiscorbutic* (prevents scurvy) factor. In 1912, Casimir Funk introduced his theory that scurvy is due to the absence of an "anti-scurvy vitamine." This factor was named vitamin C in the 1920s. Albert Szent-Györgyi isolated a substance he called *hexuronic acid* because it is a six-carbon compound. Hexuronic acid was later renamed *ascorbic acid*. Albert Szent-Györgyi was awarded the 1937 Nobel Prize in medicine for the discovery of ascorbic acid (vitamin C) and its role in preventing scurvy. By 1932, vitamin C was isolated and identified, complete with photographs of the vitamin C crystals. The ascorbic acid molecule was first successfully synthesized in 1933.

The Most Popular Supplement

The Swiss pharmaceutical company Hoffman-La Roche was the first to mass-produce vitamin C. Today, many thousands of tons of ascorbic acid are produced synthetically from glucose. The glucose is often derived from corn syrup.

Many consumers believe that vitamin C is beneficial, as about 30 percent of the U.S. adult population takes supplemental vitamin C. Currently, vitamin C is the most widely used vitamin supplement in the world.

Vitamin C is also used extensively to preserve food. L-ascorbic acid and its fatty acid esters are food additives used as browning inhibitors, antioxidants, flavor stabilizers, dough modifiers, and color stabilizers. *Ascorbyl palmitate*, a form of vitamin C, is sometimes used in antioxidant preparations because of its greater solubility in fats and oils.

Biosynthesis of Vitamin C

The vast majority of animals synthesize vitamin C. It is not a vitamin for them as they do not need to obtain it from food. The few animals that cannot synthesize vitamin C in their bodies include apes, humans, guinea pigs, one type of bird, one type of bat, and one type of fish. All other animals synthesize vitamin C in a four-step process.

The first step in synthesizing vitamin C starts with *glucuronate*, which is made from blood sugar (glucose), as seen in Figure 2-1. In the second step, glucuronate is converted to *L-gluconate*. The third step changes the L-gluconate to *L-gulono-gamma-lactone*. It is interesting to note that humans can perform these three initial steps to synthesize vitamin C. Humans are lacking the enzyme *gulonolactone oxidase*, in which L-ascorbic acid is made from L-gulono-gamma-lactone. We cannot synthesize vitamin C in our bodies, and must ingest it. This fourth step is catalyzed in the liver of most animals.

The amount of vitamin C that humans would synthesize, if they had the ability, may be a clue to our needs. The rate of biosynthesis of vitamin C in those species capable of producing the vitamin varies considerably between species.

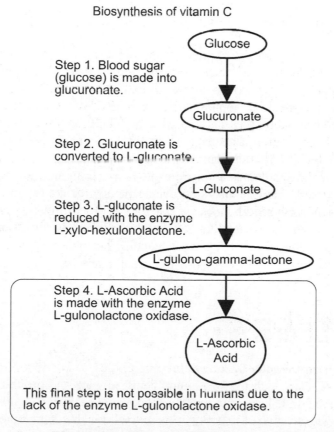

Figure 2-1   Biosynthesis of vitamin C.

The lowest biosynthesis of vitamin C is 40 mg per kilogram of body weight daily for the dog or cat; this works out to 2800 mg daily for a person of 70 kilograms (154 pounds). The highest biosynthesis of vitamin C is in the mouse, which synthesizes 275 mg of vitamin C per kilogram of body weight daily; this works out to about 20 grams (20,000 mg) daily for a 70-kilogram person. If humans could synthesize their own vitamin C, the amount synthesized daily might be between 2800 mg and 20,000 mg.

> If humans could synthesize their own vitamin C, the amount synthesized daily might be between 2800 mg and 20,000 mg.

Collagen and Vitamin C

Collagen is a fibrous protein that provides the structure of connective tissue such as skin, arteries, tendons, bones, teeth, and cartilage. Collagen is tough and has high tensile strength. Collagen protein comprises almost half of all of the protein in the body. Collagen production, supported by vitamin C, is also important for veins, heart valves, intervertebral discs, the cornea, and the lens of the eye. Collagen is a large molecule consisting of about one thousand amino acid residues. Collagen is mostly composed of only two amino acids, glycine and hydroxyproline.

The collagen molecule has a unique triple-helix configuration with three intertwined polypeptide chains. Collagen provides the organic matrix upon which bone minerals crystallize. Collagen glues cut skin together by forming a scar. This is why surgeons will often recommend higher levels of vitamin C supplementation after surgery to support increased collagen production. The collagen protein is invaluable in artery and capillary walls to provide the strength and flexibility needed to resist cardiovascular disease.

HOW COLLAGEN IS MADE

It is a major manufacturing effort for our bodies to produce collagen. First, *procollagen* is made with the two amino acids, glycine and proline, as shown in Figure 2-2. Vitamin C is used in making procollagen. The conversion of procollagen to collagen involves a reaction (*hydroxylation*) that substitutes a hydroxyl group, OH, for a hydrogen atom, H. The proline residues at certain points in the polypeptide chains are *hydroxylated* to *hydroxyproline*. The essential amino acid

Biosynthesis of Collagen

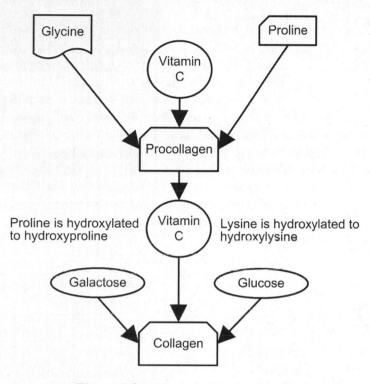

Figure 2-2   Biosynthesis of collagen.

lysine is hydroxylated to *hydroxylysine*, which is needed to permit the cross-linking of the triple helices of collagen into the fibers and networks of the tissues. One molecule of vitamin C is destroyed each time proline or lysine is hydroxylated. Lack of vitamin C can cause weak and brittle arteries and raise the risk of cardiovascular disease because of lowered collagen production.

Collagen and *keratin*, another fibrous protein, are responsible for the elasticity of skin. The degradation of collagen leads to wrinkles that can be caused by aging or smoking. These wrinkles are partly due to oxidative damage by free radicals. As we age, collagen becomes more highly cross-linked and therefore more rigid. Injected collagen is used in cosmetic surgery, especially to thicken lips.

Ferrous iron is required as a cofactor for collagen synthesis in the body. Vitamin C plays its role as an antioxidant to reduce the iron from its oxidized state. The iron can then go on as a cofactor to make more collagen.

Vitamin C as an Antioxidant

Vitamin C is one of the most important antioxidants. Vitamin C safeguards the water-soluble substances in the body from damage by *free radicals*. Free radicals are molecules with an unpaired electron—they are hungry for another electron. Free radicals are unstable and react quickly with other compounds. Normally, free radicals attack the nearest stable molecule, "stealing" its electron, as seen in Figure 2-3. This attack is known as *oxidative stress*. When the "attacked" molecule loses its electron, it can become a free radical itself, beginning a chain reaction that can continue. If this chain reaction continues, it can disrupt a living cell.

Free radicals occur normally during metabolism. Also, the body's immune system purposefully creates them to neutralize viruses and bacteria. Other sources of free radicals and other oxidative stresses include environmental factors such as pollution, hard radiation, cigarette smoke, and certain pesticides.

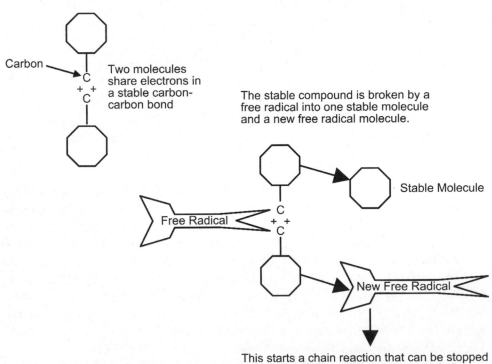

Figure 2-3   Free radicals in action.

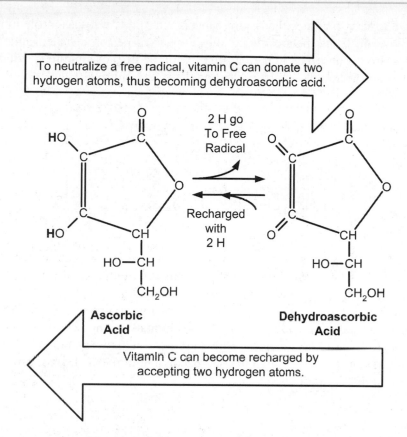

To neutralize a free radical, vitamin C can donate two hydrogen atoms, thus becoming dehydroascorbic acid.

2 H go To Free Radical

Recharged with 2 H

Ascorbic Acid

Dehydroascorbic Acid

Vitamin C can become recharged by accepting two hydrogen atoms.

Figure 2-4   Vitamin C as an antioxidant.

Vitamin C is unusual in that it can neutralize a free radical without becoming a free radical itself. As seen in Figure 2-4, vitamin C can donate one or two hydrogen atoms to neutralize a free radical. The electrons from the hydrogen atoms neutralize the free radicals to prevent a free radical chain reaction. After donating two hydrogen atoms with their electrons, vitamin C can normally be reactivated with the addition of two hydrogen atoms. Vitamin C with its two hydrogen atoms is called *ascorbic acid* and is in a state of readiness to perform antioxidant actions. Vitamin C is called *dehydroascorbic acid* when it has lost its two hydrogen atoms. Dehydroascorbic acid needs to be recharged before acting as an antioxidant. Vitamin C is stable both with and without its extra hydrogen atoms.

Vitamin C can protect many indispensable molecules in the body, such as proteins, fats, carbohydrates, and nucleic acids (DNA and RNA), from damage by free radicals. The protection of DNA from oxidative damage is one way that vitamin C can help reduce the risk of cancer. Vitamin C also has a role in regenerating vitamin E and beta-carotene after they have performed their antioxidant functions. Iron is absorbed in the intestines with the help of vitamin C. Vitamin C assists the absorption of iron by protecting the iron from oxidation.

> ### Antioxidant Roles of Vitamin C
>
> Protects Protein and DNA.
> Regenerates vitamin E and beta-carotene.
> Aids absorption of iron.
> Inhibits the formation of nitrosamines.

Nitrates are present in many common foods. These nitrates can be transformed into cancer-causing nitrosamines in the intestines if vitamin C is lacking. There is evidence that vitamin C may inhibit the formation of cancer-causing nitrosamines from nitrate and may reduce the carcinogenicity of preformed nitrosamines. Vitamin C neutralizes the nitrosamines by working as an antioxidant and donating electrons. Meats are often cured with nitrates and may contain nitrosamines. Nitrosamine content can be high in fried bacon, cured meats, beer, tobacco products, and nonfat dry milk.

Vitamin C, Infections, and the Common Cold

The idea that vitamin C supplementation might be of benefit against colds achieved wide popularity in the 1970s when Linus Pauling wrote a best-selling book titled *Vitamin C and the Common Cold.* Linus Pauling was a prominent chemist who won two Nobel prizes. Whether or not vitamin C helps prevent or cure the common cold remains a matter of great controversy. While supplementary vitamin C has not been confirmed to lower the incidence of colds, vitamin C consumed in fruits and vegetables is correlated with a lower incidence of colds. There are many other excellent nutrients in fresh produce that may contribute to fewer colds. Vitamin C taken as a supplement in doses totaling one gram or more daily has been correlated with lessened severity and slightly shorter duration of colds in some studies.

To fight infections, immune system phagocytes release oxidizing agents to kill viruses and bacteria. These oxidizing agents can also be harmful to our own cells.

Vitamin C, in its role as an antioxidant, protects our cells against the free radicals released by phagocytes. Histamine is produced by immune cells and can cause inflammations such as a stuffy nose. The antihistamine effects of vitamin C can help with the symptoms of a stuffy nose.

How Vitamin C May Help with Colds

By reducing damage from phagocyte-released free radicals.

With an antihistamine effect.

By keeping phagocytes and leukocytes "charged" with vitamin C.

By increasing the production of interferon.

By reducing the likelihood of a cold progressing to pneumonia.

More vitamin C is needed during colds.

It is also interesting that vitamin C is found in high concentrations in phagocytes and lymphocytes, indicating that vitamin C may have important functional roles in these immune system cells. Vitamin C increases the production of interferon in the body. Interferon inhibits viral proliferation, including colds. Lack of vitamin C has been found to raise the risk of many infections, especially pneumonia. One benefit identified in cold studies is an 85 percent reduction in pneumonia incidence with supplemental vitamin C intake.

Studies with human subjects have found decreased vitamin C levels in plasma, leukocytes, and urine during various infections including the common cold. This may indicate that more vitamin C is needed during infections. The vitamin C levels inside leukocytes can be reduced to half during a cold, but the level returns to the original level about a week after the episode. Vitamin C supplementation in high doses of six grams daily has been found to greatly reduce the decline of vitamin C in leukocytes caused by colds. Vitamin C levels are also depleted by stress, smoking, and alcoholism.

Vitamin C and Disease Prevention

Good general advice for disease prevention is to eat at least five servings of fresh fruit and vegetables daily. It is likely, but not certain, that additional supplemental vitamin C can further lower the risk of some diseases, especially if abundant fresh fruits and vegetables are not a regular part of the diet.

VITAMIN C AND HEART DISEASE

There is some controversy over whether vitamin C can reduce the risk of coronary heart disease. Some studies have found no correlation. However, large studies have shown that vitamin C intakes of 400 to 700 mg per day reduce the risk of coronary heart disease by about 25 percent. These studies followed more than a quarter of a million people for at least ten years.

> Vitamin C may lower the risk of:
>
> Coronary heart disease.
> Cancer.
> Cataracts.
> High lead levels.

VITAMIN C AND CANCER

Vitamin C is well known to reduce the risk of contracting a wide variety of cancers. In many cases, a healthy diet can provide enough vitamin C to be protective against cancer. This is the basis for the dietary guidelines endorsed by the National Cancer Institute and the U.S. Department of Agriculture, which recommend at least five servings of fruit and vegetables daily. Study results vary widely for breast cancer with some studies showing no protective effect and other studies showing a strong protective effect with intakes over 200 mg per day. Many studies have shown that higher intakes of vitamin C are associated with decreased incidence of cancers of the mouth, throat, esophagus, stomach, colon, and lungs.

VITAMIN C, CATARACTS, AND LEAD

Cataracts are a major cause of visual impairment, especially as people age. Not all studies have shown a protective effect on cataracts from vitamin C. There may be decreased severity of cataracts if vitamin C intake exceeds 300 mg daily for many years.

The heavy metal lead is associated with many health problems, especially in children. Blood levels of lead are generally lower when vitamin C intake is higher.

Other Roles of Vitamin C

Vitamin C is essential for the synthesis of carnitine from the essential amino acid lysine. Carnitine is required for transport of fatty acids into the mitochondria where the fatty acids are used for energy production, as seen in Figure 2-5. As you may recall, one of the early symptoms of vitamin C deficiency is fatigue, which may be related to the decreased synthesis of L-carnitine.

Vitamin C is needed to make the neurotransmitter *norepinephrine*. Norepinephrine is created in response to physical stress. This is one reason why more vitamin C is needed for stress. Vitamin C also acts as a cofactor for the enzyme that catalyzes the conversion of the neurotransmitter dopamine to norepinephrine.

> More vitamin C is needed in times of stress.

The adrenal glands contain more vitamin C than any other organ in the body. In response to stress, vitamin C is released along with stress hormones. Many different kinds of stress cause the release of vitamin C from the adrenals. More vitamin C is needed during extreme hot and cold temperatures and with exposure to heavy metals such as lead, cadmium, or mercury.

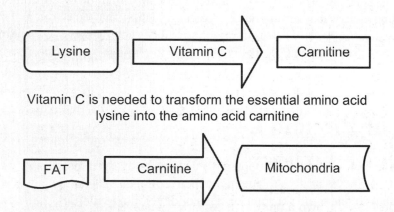

Vitamin C is needed to transform the essential amino acid lysine into the amino acid carnitine

Carnitine is needed to transport fatty acids to the mitochondria for energy production

Figure 2-5 Vitamin C and carnitine are needed for burning fat.

Certain medications, if taken for extended periods, can increase the amount of vitamin C required. Some examples of these drugs are aspirin, oral contraceptives, and barbiturates. For example, two aspirin tablets taken every six hours for a week have been reported to lower the amount of vitamin C in white blood cells by half.

Vitamin C has been used under medical supervision to treat cancer with variable success. High intravenous doses have sometimes resulted in lessening of pain and extended life. This treatment is very controversial.

Vitamin C is needed to catalyze enzymatic reactions that activate hormones such as *oxytocin*. Oxytocin is needed by women during labor to stimulate contractions. Oxytocin also aids in the release of breast milk.

> Vitamin C works in two ways to lower cholesterol.
> Vitamin C is needed to change cholesterol into bile in the liver.
> Vitamin C helps contract the gallbladder to release bile.

Vitamin C is needed for the transformation of cholesterol into bile acids and the elimination of bile acids through the gallbladder. In vitamin C deficiency, the breakdown of cholesterol is slowed, resulting in an accumulation of cholesterol in the liver. This can lead to high blood cholesterol and the formation of gallstones. Vitamin C is needed to catalyze another enzymatic reaction that activates a polypeptide hormone, *cholecystokinin*. This hormone stimulates contraction of the gallbladder to release bile. So, vitamin C works in two ways to eliminate excess cholesterol.

Deficiency of Vitamin C

Deficiency of vitamin C is apparent in many body systems. Blood vessels are dependent on vitamin C to maintain their collagen. Bleeding gums and small hemorrhages under the skin can be signs of developing vitamin C deficiency. With severe deficiency, the muscles deteriorate, including the heart muscle. Wounds fail to heal and teeth become loosened because collagen cannot be formed. Infections are common. Death from scurvy can be from internal hemorrhaging. Scurvy is easily reversed with fresh fruit and vegetables.

Vitamin C Food Sources

Vitamin C is commonly found in food as L-ascorbic acid or dehydroascorbic acid. Vitamin C is most abundant in vegetables and in fruit, especially citrus fruit; please refer to Graph 2-1. Parsley and peppers are rich in vitamin C. Potatoes are a good source of vitamin C. The best source of vitamin C is fresh, uncooked produce. Vitamin C is easily absorbed by active transport from the intestines. Vitamin C is rapidly depleted in the body and stores are critically low after a month without vitamin C.

The newest RDAs for vitamin C are 75 mg for women, 90 mg for men, 120 mg for breastfeeding women, and 125 mg for male smokers. Five servings of fresh fruits and vegetables provide approximately 200 mg of vitamin C. Many Americans do not eat enough fresh produce to meet these RDAs. Scurvy will not normally develop unless vitamin C intake falls below 10 mg per day for a month or more. The minimum vitamin C intake to maintain normal metabolism is 30 mg daily. Even though clinical scurvy may not develop, many other important body functions may be limited if vitamin C intake is low.

Optimal amounts for disease prevention of the ascorbated form of vitamin C range from 200 mg to 2 grams daily. The ascorbic acid form should not be taken in excess of 100 mg per dose because of the possibility of digestive irritation. Please refer to Table 2-1 for RDAs and adequate daily intake levels (AI) for vitamin C for all ages.

Summary for Ascorbic Acid—Vitamin C

Main function: Collagen formation and antioxidant.
RDA: 75 mg to 125 mg for adults.
No toxicity reported.
Mild digestive irritation possible over 100 mg with the
   ascorbic acid form.
Tolerable upper intake level is set at 2000 mg daily.
Deficiency causing scurvy is rare. Many people do not
   achieve the RDAs.
Deficiency disease: scurvy.
Healthy food sources: found in fresh fruits and vegetables.
Degradation: leached by cooking water, reduced by heat,
   light, light, oxygen, and food processing.

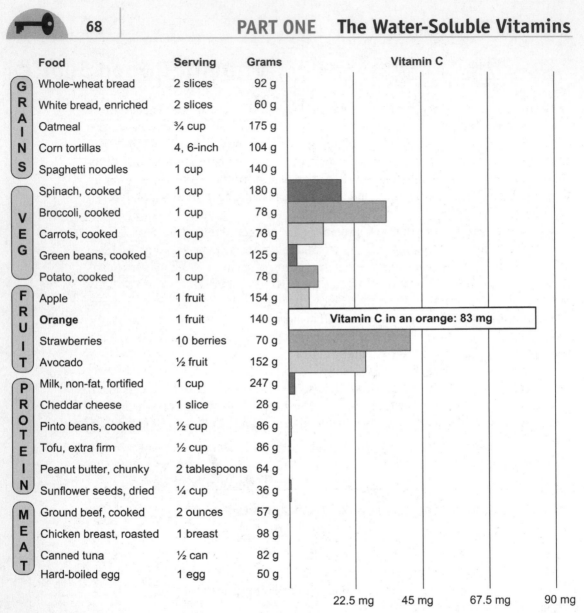

Food	Serving	Grams	Vitamin C
GRAINS			
Whole-wheat bread	2 slices	92 g	
White bread, enriched	2 slices	60 g	
Oatmeal	¾ cup	175 g	
Corn tortillas	4, 6-inch	104 g	
Spaghetti noodles	1 cup	140 g	
VEG			
Spinach, cooked	1 cup	180 g	
Broccoli, cooked	1 cup	78 g	
Carrots, cooked	1 cup	78 g	
Green beans, cooked	1 cup	125 g	
Potato, cooked	1 cup	78 g	
FRUIT			
Apple	1 fruit	154 g	
Orange	1 fruit	140 g	Vitamin C in an orange: 83 mg
Strawberries	10 berries	70 g	
Avocado	½ fruit	152 g	
PROTEIN			
Milk, non-fat, fortified	1 cup	247 g	
Cheddar cheese	1 slice	28 g	
Pinto beans, cooked	½ cup	86 g	
Tofu, extra firm	½ cup	86 g	
Peanut butter, chunky	2 tablespoons	64 g	
Sunflower seeds, dried	¼ cup	36 g	
MEAT			
Ground beef, cooked	2 ounces	57 g	
Chicken breast, roasted	1 breast	98 g	
Canned tuna	½ can	82 g	
Hard-boiled egg	1 egg	50 g	

22.5 mg   45 mg   67.5 mg   90 mg

Graph 2-1 Vitamin C in some common foods.

Table 2-1   Recommended daily intake levels for vitamin C.

RDAs for Vitamin C	Age	Males mg/day	Females mg/day
Infants	0–6 months	40 (AI)	40 (AI)
Infants	7–12 months	50 (AI)	50 (AI)
Children	1–3 years	15	15
Children	4–8 years	25	25
Children	9–13 years	45	45
Adolescents	14–18 years	75	65
Adults	19 years and older	90	75
Adult smokers	19 years and older	125	110
Pregnancy	18 years & younger	—	80
Pregnancy	19 years and older	—	85
Breastfeeding	18 years & younger	—	115
Breastfeeding	19 years and older	—	120

Vitamin C is easily leached out into water during cooking. It is also lost from heat, exposure to light, and exposure to oxygen. Prolonged storage and food processing reduce the amount of vitamin C in food, as shown in Figure 2-6.

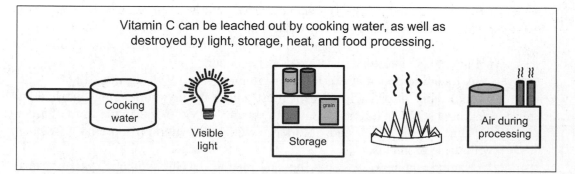

Figure 2-6   Vitamin C is easily destroyed during storage and processing.

Supplemental Forms of Vitamin C

Vitamin C is rapidly absorbed and rapidly eliminated from the body. When using supplements containing vitamin C, it is best if the vitamin C is released slowly, over a period of time. This reduces the rebound effect, where less vitamin C is available than normal after a supplementary dose has been used up. Taking vitamin C supplements throughout the day is another way of ensuring availability. Taking a timed-release form of vitamin C before going to bed can help keep it available all night.

> ### Guidelines for Vitamin C Supplements
>
> Use timed-release tablets or periodic dosing.
> The ascorbated form is less acidic and better
>   transported.
> Include bioflavonoids to offset possible
>   capillary fragility.

Vitamin C in the form of ascorbic acid has a pH of 2.8, which is slightly less acidic than lemon juice. This acidity is not usually a problem if less than 100 mg are taken in any one dose. However, if doses are higher, such as one gram or above, the acidity may irritate the intestines or urinary tract, causing mild discomfort or diarrhea. To avoid this acidity, vitamin C supplements can be *ascorbated* with a mineral.

Ascorbation is a process where an acidic vitamin is combined chemically with an alkaline mineral. *Calcium ascorbate* and *sodium ascorbate* are the most common forms, although many other minerals can be ascorbated. The ascorbated vitamin C is easier for the body to transport. Ascorbated forms of vitamin C are pH neutral, so intestinal irritation does not occur.

A tolerable upper intake level for vitamin C is set at 2 grams (2,000 milligrams) daily in order to prevent most adults from experiencing diarrhea and gastrointestinal disturbances due to the acidity of vitamin C in the form of ascorbic acid. On the other hand, ascorbates, with their neutral pH, are not associated with these problems.

Natural vitamin C in food is the same chemical as the synthetic L-ascorbic acid found in supplements. The natural ascorbic acid in food is digested and absorbed slowly, providing a timed-release effect. Also, because of natural buffering, many foods that contain ascorbic acid do not cause irritation of the intestines. One exception is citrus juice, which, in excess, can irritate the intestines.

> Vitamin C in food has several advantages over synthetic ascorbic acid.

Natural ascorbic acid found in food is also commonly accompanied by bioflavonoids. The white partitions of citrus fruits are rich in bioflavonoids. Bioflavonoids have their own powerful antioxidant effects. One of the effects of bioflavonoids is to decrease capillary fragility. Since ascorbic acid slightly increases capillary fragility, this action of the bioflavonoids offsets this tendency. Natural ascorbic acid in food may also ascorbate with minerals in the food, thereby easing absorption and transport.

Some vitamin C supplements contain small amounts of the vitamin C metabolite dehydroascorbate (oxidized ascorbic acid) and other vitamin C metabolites. Absorption has not been shown to be higher with these additions. Other supplements contain a fat-soluble form of vitamin C known as ascorbyl palmitate. The ascorbic acid is combined with palmitic acid and becomes fat soluble. Ascorbyl palmitate is often used in cosmetic creams as a fat-soluble antioxidant. Taken orally, ascorbyl palmitate is broken down to ascorbic acid and palmitic acid before absorption.

Toxicity of Vitamin C

Vitamin C is one of the least toxic substances used in supplements. Even in huge doses of 10 to 20 grams daily, no health problems or side effects were noted, other than gastrointestinal disturbances if ascorbates were not used. There has not been reliable data to show that vitamin C has a clear relationship with kidney stone formation in the human body. This is in spite of the fact that excess vitamin C in the blood does break down to oxalic acid and is eliminated through the kidneys.

In test tube experiments, vitamin C can interact with some free metal ions, such as iron, to produce potentially damaging free radicals. However, free metal ions are not generally found in the body. Supplemental vitamin C has not been found to promote these free radicals inside a human body.

Generous amounts of vitamin C keep us healthy and protect us from illness in many ways.

Quiz

Refer to the text in this chapter if necessary. A good score is at least 8 correct answers out of these 10 questions. The answers are listed in the back of this book.

1. The main role of vitamin C is:
   (a) Colloidal production.
   (b) Energy metabolism.
   (c) Collagen production.
   (d) Preventing the common cold.

2. The cure for scurvy is:
   (a) Citrus fruit.
   (b) Vitamin C tablets.
   (c) Fresh vegetables.
   (d) All of the above.

3. Humans can synthesize vitamin C:
   (a) In their livers.
   (b) In the adrenal glands.
   (c) From sunlight.
   (d) None of the above.

4. Vitamin C is essential for the synthesis of:
   (a) Carnitine.
   (b) Proline.
   (c) Lysine.
   (d) Arginine.

5. The organ with the highest levels of vitamin C is:
   (a) Heart.
   (b) Adrenals.
   (c) Pancreas.
   (d) Bones.

6. Vitamin C:
   (a) Raises blood cholesterol.
   (b) Has no effect on blood cholesterol.
   (c) Lowers blood cholesterol.
   (d) Raises homocysteine.

7. Vitamin C:
   (a) Is found in fish and chicken.
   (b) Does not leach out during cooking.
   (c) Is rapidly depleted if not present in the diet.
   (d) Is found in dairy products.

8. Supplemental vitamin C is best if:
   (a) It is ascorbated and timed-release.
   (b) Taken as ascorbic acid powder.
   (c) Taken without bioflavonoids.
   (d) All of the above.

9. The RDA for adults for vitamin C is:
   (a) One to two grams per day.
   (b) 45 to 60 mg per day.
   (c) 75 to 125 mg per day.
   (d) 95 to 150 mg per day.

10. Supplementary forms of vitamin C are:
    (a) Ascorbyl palmitate.
    (b) Mineral ascorbates.
    (c) Dehydroascorbate.
    (d) All of the above.

Test: Part One

Do not refer to the text when taking this test. A good score is at least 75 correct. Answers are in the back of the book. It's best to have a friend check your score the first time, so you won't memorize the answers if you want to take the exam again.

1. Water-soluble vitamins:
   (a) Are made in the watery parts of the body.
   (b) Excesses are usually regulated by the kidneys.
   (c) Are not subject to leaching.
   (d) Are found in water.

2. Riboflavin is known as:
   (a) Vitamin B_1.
   (b) Vitamin B_2.
   (c) Vitamin B_3.
   (d) Vitamin B_4.

3. Microwave cooking destroys:
   (a) Pyridoxine.
   (b) Pantothenic acid.
   (c) Cobalamin.
   (d) Thiamin.

4. The only vitamin that contains cobalt is:
   (a) Biotin.
   (b) Niacin.
   (c) Thiamin.
   (d) Cobalamin.

5. A perceived deficiency of pyridoxine may also be caused by a deficiency of:
   (a) Riboflavin.
   (b) Cobalamin.
   (c) Biotin.
   (d) Pantothenic acid.

6. Our bodies make energy using:
   (a) Catabolism.
   (b) Anabolism.
   (c) Cannibalism.
   (d) Electricity.

7. B vitamins act as:
   (a) Enzymes.
   (b) Coenzymes.
   (c) Fuel.
   (d) Fat.

8. B vitamins:
   (a) Are needed for nerve conduction.
   (b) Are needed to make neurotransmitters.
   (c) Are needed to convert amino acids.
   (d) All of the above.

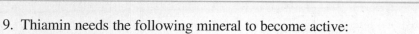

9. Thiamin needs the following mineral to become active:

   (a) Magnesium.

   (b) Manganese.

   (c) Iron.

   (d) Calcium.

10. When triple the RDA of a B vitamin is taken, they are:

   (a) Very toxic.

   (b) Mildly toxic.

   (c) A little toxic.

   (d) Not toxic.

11. One thousand milligrams is:

   (a) One gram.

   (b) One microgram.

   (c) One-tenth of a gram.

   (d) One-hundredth of a gram.

12. Thiamin deficiency can cause:

   (a) Scurvy.

   (b) Pellagra.

   (c) Beriberi.

   (d) Blindness.

13. Riboflavin and folic acid can be destroyed by:

   (a) Light.

   (b) Cool temperatures.

   (c) Bacteria and viruses.

   (d) Microwave cooking.

14. Which vitamin turns the color of urine yellow?

   (a) Thiamin.

   (b) Riboflavin.

   (c) Niacin.

   (d) Pyridoxine.

15. Biotin deficiency is:

    (a) Common.

    (b) Uncommon.

    (c) Rare.

    (d) Very rare.

16. Adequate intake of pantothenic acid is set at:

    (a) 1.2 mg.

    (b) 1.2 mcg.

    (c) 1.2 g.

    (d) 5 mg.

17. Excessive folic acid intake can mask a deficiency of:

    (a) Cobalamin.

    (b) Niacin.

    (c) Riboflavin.

    (d) Thiamin.

18. Synthetic folic acid is about:

    (a) Half as strong as natural folate.

    (b) Twice as strong as natural folate.

    (c) The same strength as natural folate.

    (d) Worthless compared to natural folate.

19. Homocysteine, which is reduced by several vitamins, is:

    (a) A needed nutrient.

    (b) Almost a vitamin.

    (c) An unhealthy blood constituent.

    (d) An essential amino acid.

20. The adult RDA for cobalamin is:

    (a) 2.4 mg.

    (b) 2.4 mcg.

    (c) 2.4 g.

    (d) 24 mcg.

21. The mitochondria:

    (a) Eliminates excess B vitamins.

    (b) Eliminates excess vitamin C.

    (c) Is a harmful metabolic waste.

    (d) Is where the B vitamins help to make energy.

22. Pyruvate is:

    (a) Normally made from glucose.

    (b) An intermediary in energy production.

    (c) Made into acetyl coenzyme A.

    (d) All of the above.

23. The electron transport chain is:

    (a) Used to energize ATP.

    (b) Used to contract nerves.

    (c) Used to produce vitamins.

    (d) A powerful antioxidant.

24. Vitamin C is needed for:

    (a) Collagen formation.

    (b) Antioxidant activity.

    (c) Both (a) and (b).

    (d) Neither (a) nor (b).

25. To consume the RDA of vitamin C, one must:

    (a) Eat an apple a day.

    (b) Eat five servings of fruit and vegetables.

    (c) Eat plenty of whole grains and meat.

    (d) Eat five servings of dairy products and whole grains.

PART TWO

The Fat-Soluble Vitamins

Introduction to the Fat-Soluble Vitamins

The fat-soluble vitamins are vitamins A, D, E, and K. Fat-soluble vitamins are found in the liver and the fatty tissues of the body where they are stored and used. Fat-soluble vitamins require bile in order to be absorbed into the lymph system from the intestines. One of the fat-soluble vitamins, Vitamin E, is an important antioxidant that protects cell membranes and artery walls.

Vitamin A comes from two different sources. Provitamin A is found in vegetables and fruits and is known as beta-carotene. Beta-carotene is always non-toxic in foods, even if you consume a lot of vegetables and fruits. Beta-carotene functions as an antioxidant and can be converted to the other forms of vitamin A. The other dietary source of vitamin A is found mainly in animal livers. Vitamin A from animal products and supplements has the potential to be toxic. Vitamin A has many functions, one of which is enhancing vision in low light.

Vitamin D is important for maintaining strong bones. Vitamin D is normally made in the skin with the help of sunlight. Supplementation may be necessary for people who do not get much sunlight. This vitamin has few natural dietary sources. Vitamin D is added to milk and is found in the oils of a few fish. Vitamin D is never toxic when made from sunlight in the skin, but it can be toxic when taken as fish liver oil or in supplements.

Vitamin E is a family of related compounds. Vitamin E is mainly found in almonds, sunflower seeds, and cold-pressed oils. Vitamin E serves as a powerful antioxidant, protecting fatty areas in the body. Vitamin E is non-toxic, although huge doses can encourage a tendency toward bleeding.

Vitamin K is needed for normal blood clotting. Vitamin K is found in abundance in green leafy vegetables. Vitamin K is non-toxic in food and supplements. Unlike the other fat-soluble vitamins, vitamin K is not stored in the body.

CHAPTER 3

Vitamin A
The Night Sight Vitamin

As long ago as 300 B.C., the Hippocratic School of Medicine recommended liver (rich in vitamin A) for children with night blindness or infections. Vitamin A was the first vitamin discovered. In 1907, the fat-soluble vitamin A was found necessary for growth. In 1930 it was learned that there were two related forms—*beta-carotene* and a fat-soluble Vitamin A. Vitamin A was first synthesized in 1947. Most of the vitamin A in the body is stored in the liver in the form of *retinyl palmitate*.

The Forms of Vitamin A

Vitamin A is a family of compounds with similar structures called *retinoids*. In plant-based foods, vitamin A is found in the form of *provitamin A*, principally *beta-carotene*. These plant-based carotenes are known as provitamin A because some of them can be sliced apart to become the other forms of vitamin A. Foods derived from animals or animal products contain a different form of vitamin A called *retinyl esters*. The retinyl esters can also be converted to the other forms of vitamin A in the body.

> ### Forms of Vitamin A
>
> **Carotenes** such as **beta-carotene** are from plant foods and are antioxidants.
> **Retinyl esters** come from animal foods.
> **Retinol** supports healthy skin.
> **Retinal** is needed for vision.
> **Retinoic acid** supports skin and epithelial tissue.
> **Retinyl palmitate** is the storage form inside the human liver.

Inside the body vitamin A is found in five forms: retinol, retinal, retinoic acid, retinyl palmitate, and beta-carotene, as seen in Figure 3-1. Each of these forms of vitamin A performs functions that the others cannot. *Retinol* is the major form of vitamin A for transport in the body. Retinol is the alcohol form of vitamin A.

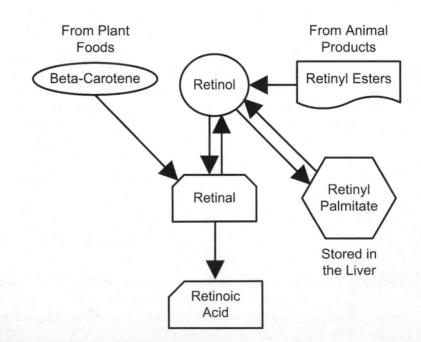

Figure 3-1   The forms of vitamin A.

Retinol is required to maintain the integrity and immune function of the skin and mucous membranes. Retinol can be converted to three of the other forms inside the body.

Retinal is the form of vitamin A famous for working in the rod cells of the eyes to enhance night vision. Retinal is the oxidized form of retinol. Retinal is a needed intermediary in the conversion of retinol to another active form of vitamin A, *retinoic acid*. Retinal can be oxidized to produce this third form of vitamin A, retinoic acid. Once retinal is oxidized into retinoic acid, it cannot be changed back to retinal again, as previously seen in Figure 3-1.

Retinoic acid acts as a hormone, affecting genes in the nucleus of the cell. Retinoic acid influences gene transcription, the expression of genes, and the synthesis of proteins. Retinoic acid regulates the developing cells for specialized uses during growth and embryonic development.

All of the forms of vitamin A are called *preformed* except provitamin A, which is also known as beta-carotene. Beta-carotene can be split into two, and each half can become a molecule of retinal, as seen in Figure 3-2.

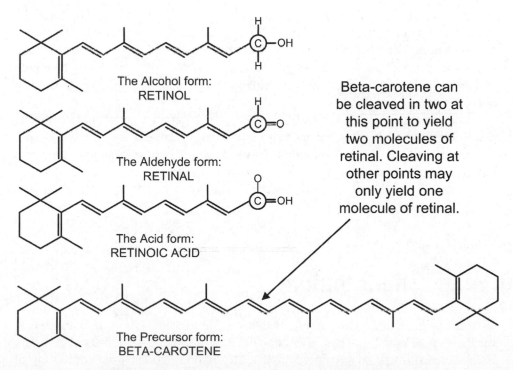

The Alcohol form:
RETINOL

The Aldehyde form:
RETINAL

The Acid form:
RETINOIC ACID

The Precursor form:
BETA-CAROTENE

Beta-carotene can be cleaved in two at this point to yield two molecules of retinal. Cleaving at other points may only yield one molecule of retinal.

Figure 3-2   Chemical structures of vitamin A forms.

Antioxidant Activity

Vitamin A has several forms that are used for vital functions. Provitamin A, *beta-carotene*, performs antioxidant functions that none of the other forms of vitamin A can achieve. In addition to its vital antioxidant functions, beta-carotene can be split apart into retinal and converted to all other forms of preformed vitamin A, as previously seen in Figures 3-1 and 3-2.

Beta-carotene is one of the most powerful antioxidants in food. Antioxidants neutralize free radicals to reduce the risk of macular degeneration, cancer, heart disease, and stroke. Some of the beta-carotene in foods and supplements can be converted into the retinal form of vitamin A. About 10 percent of the *carotenoids* (beta-carotene is one of the carotenoids) in plant foods can be converted into retinal. The remaining carotenoids may be used as antioxidants.

> Beta-carotene is abundant in yellow and orange vegetables and fruit.

The other forms of vitamin A do not exhibit antioxidant activity. The forms of vitamin A found in meat (retinyl esters), dairy products, and eggs do not possess antioxidant activity. Vitamin A supplements made without beta-carotene or other sources of antioxidants also do not possess antioxidant activity. Many supplements are made with retinyl palmitate and retinyl acetate; these forms of vitamin A are not antioxidants.

Beta-carotene is plentiful in yellow and orange vegetables and fruit. Green vegetables also are rich in beta-carotene; the colorful pigments are masked by the green chlorophyll. Some of the other carotenoids that can be converted into retinal include *alpha-carotene* and *beta-cryptoxanthin*. Some carotenoids that cannot be converted into retinal are lycopene (from tomatoes) and lutein. All carotenoids have antioxidant activity.

Vitamin A and Night Vision

Vitamin A is needed by the retina of the eye for vision. The retina is located at the back of the eye. Light passes through the lens of the eye and hits the retina. The retina converts the light into nerve impulses for interpretation by the brain. Retinol is transported by the bloodstream to the retina. In the retina, the retinol is used by the epithelial cells on the inside surface of the retina.

Retinol is stored in the retina in the form of retinyl ester until it is needed. When it is needed, the retinyl ester is hydrolyzed into retinol, as shown in Figure 3-3. The retinol is changed to a special form of retinal that is a "*cis*" isomer. Isomers are mirror images. This cis-retinal goes to the rod cells of the retina. Rod cells in the retina are responsible for vision in dim light. In the rod cells, the retinal binds to a protein called opsin. The retinal and the opsin together form the compound

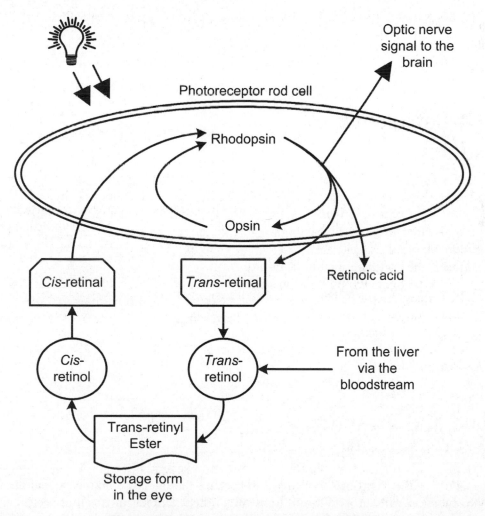

Figure 3-3   How vitamin A helps night vision.

rhodopsin, also known as *visual purple*. Rod cells with rhodopsin are able to detect tiny amounts of light; this is important for night vision.

When a photon of light hits a rod cell, cis-retinal is released and transformed to trans-retinal. This release of retinal generates an electrical signal to the optic nerve. The optic nerve sends a signal to the brain. Some of the trans-retinal becomes available for another cycle of vision. Some of the trans-retinol is converted to retinoic acid and is no longer available to bind to opsin to form rhodopsin. These losses must be replaced with vitamin A either from the diet or from vitamin A stored in the liver.

Each cell in the retina contains about 30 million molecules containing forms of vitamin A—and there are over 100 million cells in the retina. If enough retinol is not available to the retina, this can result in night blindness.

Vitamin A Deficiency and Blindness

Severe vitamin A deficiency is one of the leading causes of blindness in children. This type of blindness is not related to night vision. Over half of a million children lose their sight each year from severe vitamin A deficiency. This preventable childhood blindness results from a lack of vitamin A in the cornea of the eye. The cornea is the transparent outer layer of the eye.

This type of childhood blindness is known as *xerophthalmia*. In the first stage of xeropthalmia, the cornea becomes hard and dry, a condition known as *xerosis*. Xerosis can progress to a softening of the cornea that can lead to irreversible blindness.

Infections and Vitamin A

Vitamin A is required by the immune system in several different ways. Vitamin A is called the "anti-infective" vitamin because the human immune system cannot function without it. With millions of children deficient in vitamin A, deaths from common childhood infections such as pneumonia and measles can be greatly reduced with better food or vitamin A supplementation. While childhood vitamin A deficiency is rare in developed countries, it is widespread in southeast Asia and Africa.

VITAMIN A AND MUCOUS MEMBRANES

One of the first lines of defense against infection is the mucous membranes. These mucous membranes line the digestive tract, the lungs and sinuses, the vagina, the eyes, and the urinary tract. Vitamin A is required to maintain the integrity of these vital barriers to infection.

Without vitamin A, certain cells in the mucous membranes, *goblet cells*, become fewer in number. Goblet cells are needed to produce mucus, which is necessary for all of the mucous membranes. Mucus forms a chemical barrier to gastric acid in the stomach. Mucus also helps eliminate contaminants in the lungs. Mucus forms an important first barrier to invasive microorganisms. With fewer goblet cells, less mucus is produced, as seen in Figure 3-4. This is a problem in the intestines because

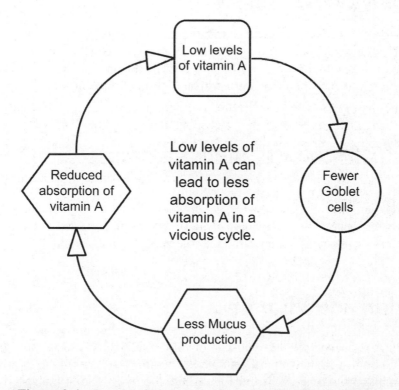

Figure 3-4   Levels of vitamin A must be maintained for proper absorption of vitamin A.

the decreased amount of mucus leads to decreased absorption of nutrients. This leads to diminished vitamin A absorption—leading to a vicious cycle.

How Vitamin A Reduces Risk of Infections

Vitamin A strengthens mucous membranes.
Vitamin A increases mucous secretion.
Vitamin A keeps skin flexible.
Vitamin A is needed in the development of lymphocytes.
Vitamin A is needed for the regulation of the immune system.

VITAMIN A KEEPS SKIN SOFT

When vitamin A is deficient, the skin becomes drier, and can become rough and scaly. This hardening and drying is related to an increase of *keratin*, a hard and inflexible protein found in fingernails. This hardened skin is less effective as a barrier to infection. This hardening effect from deficient vitamin A is also a problem with the mucous membranes throughout the body.

VITAMIN A INCREASES IMMUNE RESPONSE

Vitamin A has other roles in resistance to infection. Vitamin A plays a central role in the development of lymphocytes, white blood cells that play critical roles in the immune response. Also, activation of the major regulatory cells of the immune system, T-lymphocytes, requires the retinoic acid form of vitamin A.

Cell Formation and Vitamin A

Vitamin A is needed for the synthesis of important proteins used throughout the body. Vitamin A in the form of retinoic acid can act as a hormone to affect how genes make protein, as shown in Figure 3-5. Retinoic acid is transported to the cells bound to special proteins, *cytoplasmic retinoic acid-binding proteins*. Once inside the nucleus of the cell, retinoic acid binds to *special receptor proteins*. Here

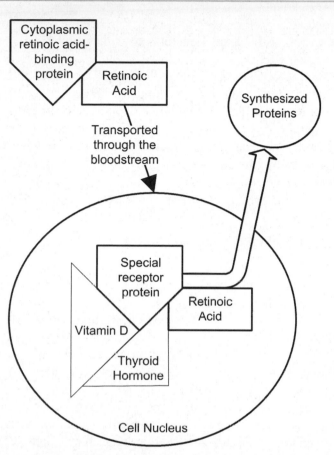

Figure 3-5 Vitamin A regulates protein production in the nucleus of the cell.

in the nucleus of the cell, retinoic acid affects gene transcription that enables synthesis of certain proteins. Thyroid hormone and vitamin D interact with retinoic acid in the nucleus of the cell to influence protein synthesis.

By regulating which genes are expressed, retinoic acid plays a major role in deciding which type of specialized cell a developing cell will become. This is especially important in embryonic development. This is also important in rapidly-developing cells such as epithelial cells in skin and mucous membranes. This change of a cell to a more specialized cell is called *differentiation*. Many of the important effects attributed to vitamin A appear to result from its role in cellular differentiation.

Vitamin A in the form of retinoic acid is essential for fetal development. Retinoic acid is needed for the formation of the heart, eyes, limbs, and ears of the growing fetus. Not only that, but retinoic acid is used to regulate the expression of the gene for *growth hormone*. Without sufficient retinoic acid, growth hormone may be limited in the developing fetus.

Millions of red blood cells are made in the body every second. Red blood cells are derived from precursor cells called stem cells. These stem cells are dependent on vitamin A for normal differentiation into red blood cells. In addition, vitamin A facilitates the mobilization of iron from storage areas to the developing red blood cell for integration into hemoglobin. Hemoglobin is the oxygen carrier in red blood cells. Now that we have seen how vitally important vitamin A is, let us look at the food sources.

Sources of Vitamin A

Certain vegetables and fruits have abundant provitamin A and others have very little, as seen in Graph 3-1. Spinach has about six times as much as broccoli. Broccoli has about ten times as much as celery and cabbage. Colored fruits and vegetables, especially the yellow and orange ones, are generally high in provitamin A. Meats are generally low in vitamin A, except for liver which is excessively high. Dairy products contain a medium amount of vitamin A.

The RDA is based on the amount needed to ensure adequate stores of vitamin A in the body to support pregnancy, gene expression, immune function, and vision. Different dietary and supplemental sources of vitamin A have different potencies. For many years vitamin A activity was measured in International Units (IU) The current international standard of measure for vitamin A activity, however, is *retinol activity equivalency* (RAE). The RAE uses a microgram (mcg) of retinol as a standard to measure the potency of the forms of vitamin A. Three mcg of RAE is equivalent to ten International Units. The RDA for an adult man is 3000 IU, which is equivalent to 900 mcg of RAE. For adult women, the RDA is 2333 IU, which is equivalent to 700 mcg of RAE. Please refer to Table 3-1 for more complete RDA information.

The best dietary supplement of vitamin A consists of generous quantities of fresh fruits and vegetables. Supplemental forms of vitamin A such as retinyl palmitate and retinyl acetate convert readily into retinol and their potency is listed on the packaging.

With supplemental beta-carotene, two mcg of beta-carotene are needed by the body for conversion, resulting in one mcg of retinol. This is known as a RAE ratio of 2:1. For beta-carotene in food, the RAE ratio is 12:1, so 12 mcg of beta-carotene in food is needed to provide one mcg of RAE. With *alpha-carotene* and *beta-cryptoxanthin*, the ratio is 24:1, so more is needed to provide the same retinol

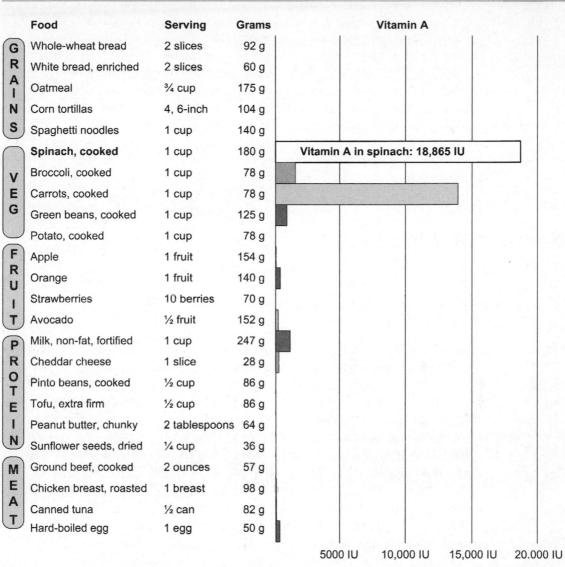

Food	Serving	Grams	Vitamin A
GRAINS			
Whole-wheat bread	2 slices	92 g	
White bread, enriched	2 slices	60 g	
Oatmeal	¾ cup	175 g	
Corn tortillas	4, 6-inch	104 g	
Spaghetti noodles	1 cup	140 g	
VEG			
Spinach, cooked	1 cup	180 g	Vitamin A in spinach: 18,865 IU
Broccoli, cooked	1 cup	78 g	
Carrots, cooked	1 cup	78 g	
Green beans, cooked	1 cup	125 g	
Potato, cooked	1 cup	78 g	
FRUIT			
Apple	1 fruit	154 g	
Orange	1 fruit	140 g	
Strawberries	10 berries	70 g	
Avocado	½ fruit	152 g	
PROTEIN			
Milk, non-fat, fortified	1 cup	247 g	
Cheddar cheese	1 slice	28 g	
Pinto beans, cooked	½ cup	86 g	
Tofu, extra firm	½ cup	86 g	
Peanut butter, chunky	2 tablespoons	64 g	
Sunflower seeds, dried	¼ cup	36 g	
MEAT			
Ground beef, cooked	2 ounces	57 g	
Chicken breast, roasted	1 breast	98 g	
Canned tuna	½ can	82 g	
Hard-boiled egg	1 egg	50 g	

5000 IU   10,000 IU   15,000 IU   20.000 IU

Graph 3-1   Vitamin A content of some common foods.

Table 3-1   RDAs for vitamin A for all ages.

RDAs for Vitamin A	Age	Males mcg/day (RAE)	Females mcg/day (RAE)
Infants	0–6 months	400 (1333 IU)	400 (1333 IU)
Infants	7–12 months	500 (1667 IU)	500 (1667 IU)
Children	1–3 years	300 (1000 IU)	300 (1000 IU)
Children	4–8 years	400 (1333 IU)	400 (1333 IU)
Children	9–13 years	600 (2000 IU)	600 (2000 IU)
Adolescents	14–18 years	900 (3000 IU)	700 (2333 IU)
Adults	19 years and older	900 (3000 IU)	700 (2333 IU)
Pregnancy	18 years & younger	—	750 (2500 IU)
Pregnancy	19 years and older	—	770 (2567 IU)
Breastfeeding	18 years & younger	—	1,200 (4000 IU)
Breastfeeding	19 years and older	—	1,300 (4333 IU)

equivalency. Still, the highest amounts of vitamin A activity (outside of liver) are found in fruits and vegetables.

The retinol activity equivalency is not relevant to the antioxidant activity of provitamin A carotenes. The carotenes that are not converted to retinol are still valuable for their antioxidant activity. The RDA for vitamin A was not intended to provide enough provitamin A for antioxidant use. There is general agreement that a diet rich in fruits and vegetables provides abundant vitamin A to satisfy all needs. As mentioned, the yellow, orange, and green fruits and vegetables are richest in provitamin A.

Toxicity of Vitamin A

Vitamin A toxicity is relatively rare. Vitamin A toxicity is called *hypervitaminosis A*. Toxicity is not caused by provitamin A carotenoids such as beta-carotene. Hypervitaminosis A is caused by over consumption of the kind of preformed vitamin A found in animal products and many supplements. This preformed vitamin A is absorbed quickly and it is slowly eliminated. Toxicity can result from long-term low intakes or from short-term high intakes of preformed vitamin A.

Vitamin A is transported through the blood bound to protein. Toxicity begins to develop when the binding proteins are all full and free retinol starts to damage cells. Children are more vulnerable to hypervitaminosis A than adults.

Normally, toxicity is associated with long-term consumption of vitamin A in excess of 10 times the RDA (8,000 to 10,000 mcg/day or 25,000 to 33,000 IU/day). People who may be susceptible to toxicity at lower doses include the elderly, pregnant women, and alcoholics. A tolerable upper intake level UL has been set at 3,000 mcg (10,000 IU)/day of preformed vitamin A for adults and less for children. It is possible to receive excessive vitamin A from the diet. Liver is dangerously high in vitamin A. A diet consisting largely of animal products can cause hypervitaminosis A.

BETA-CAROTENE IS NON-TOXIC

No tolerable upper intake level for beta-carotene has been set because it is non-toxic. Beta-carotene is stored in fat just under the skin. In very rare instances, huge overdoses of concentrated beta-carotene can cause a slight yellowing of the skin. This is not harmful and quickly goes away. It is interesting to note that the type of beta-carotene used in supplements may be harmful in large doses, especially for people who drink alcohol and smoke cigarettes.

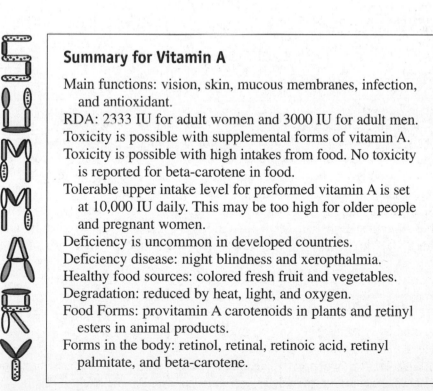

Summary for Vitamin A

Main functions: vision, skin, mucous membranes, infection, and antioxidant.

RDA: 2333 IU for adult women and 3000 IU for adult men.

Toxicity is possible with supplemental forms of vitamin A.

Toxicity is possible with high intakes from food. No toxicity is reported for beta-carotene in food.

Tolerable upper intake level for preformed vitamin A is set at 10,000 IU daily. This may be too high for older people and pregnant women.

Deficiency is uncommon in developed countries.

Deficiency disease: night blindness and xeropthalmia.

Healthy food sources: colored fresh fruit and vegetables.

Degradation: reduced by heat, light, and oxygen.

Food Forms: provitamin A carotenoids in plants and retinyl esters in animal products.

Forms in the body: retinol, retinal, retinoic acid, retinyl palmitate, and beta-carotene.

CERTAIN TYPES OF VITAMIN A CAN CAUSE BIRTH DEFECTS

Excess preformed vitamin A from animal products or supplements taken during pregnancy is known to cause birth defects. Provitamin A carotenes are not known to cause birth defects. Intake levels by pregnant women below 10,000 IU daily have not been associated with an increase of birth defects. However, many foods are fortified with preformed vitamin A that can add to the other dietary sources of vitamin A. For this reason, pregnant women are advised to keep their preformed vitamin A levels below 5000 IU daily, for example, from prenatal vitamins.

To avoid any possibility of increasing birth defects, a diet rich in vegetables and fruit is an alternative to preformed vitamin A sources. One medium-sized carrot provides a safe alternative vitamin A source and contains about 16,000 IU.

Osteoporosis and Vitamin A

In older men and women, long-term intakes of preformed vitamin A can be associated with increased risk of osteoporotic fracture and decreased bone mineral density. Levels of only 5000 IU (1,500 mcg) are enough to increase risk. This is well below the upper limit set at 10,000 IU (3000 mcg) per day. Only high intakes of preformed vitamin A, not beta-carotene, are associated with any increased adverse effects on bone health.

Older men and women may want to limit their supplemental vitamin A intake or take only the beta-carotene form of vitamin A. Many fortified foods such as cereal contain significant levels of preformed vitamin A. The vitamin A in fortified foods should be added to the vitamin A in any supplements to find the total intake. On the other hand, low levels of vitamin A can adversely affect bone mineral density. In older people, an intake of preformed vitamin A close to the RDA is safest.

The best way to assure safe levels of vitamin A is to eat plenty of fruits and vegetables and, if supplements are needed, to use the beta-carotene form.

Quiz

Refer to the text in this chapter if necessary. A good score is at least 8 correct answers out of these 10 questions. The answers are listed in the back of this book.

1. Provitamin A is known as:
   (a) Retinyl palmitate.
   (b) Retinal.
   (c) Beta-carotene.
   (d) Retinol.

2. The form of vitamin A that helps night vision is:
   (a) Carotenoids.
   (b) Retinol.
   (c) Retinyl palmitate.
   (d) Retinal.

3. The form of Vitamin A that is an antioxidant is:
   (a) Retinol.
   (b) Retinyl esters.
   (c) Retioic acid.
   (d) Beta-carotene.

4. Rhodopsin is:
   (a) Visual purple.
   (b) Found in rod cells.
   (c) Made from cis-retinal.
   (d) All of the above.

5. Goblet cells:
   (a) Make mucus.
   (b) Are eye cells.
   (c) Are important for vision.
   (d) Are found in skin.

6. Retinoic acid is transported to the cells by:
   (a) Carbohydrates.
   (b) Fats.
   (c) Proteins.
   (d) All of the above.

7. The food with potentially toxic levels of vitamin A:

    (a) Carrots.

    (b) Liver.

    (c) Cheese.

    (d) Spinach.

8. Birth defects can be caused by:

    (a) Overconsumption of carrot juice.

    (b) Vitamin A supplements at the RDA.

    (c) Provitamin A carotenoids in food.

    (d) Overconsumption of vitamin A from liver.

9. To prevent osteoporosis in older people:

    (a) Minimize vitamin A in the diet and supplements.

    (b) Take in the RDA of vitamin A.

    (c) Take in double the RDA of vitamin A.

    (d) Supplement with the maximum upper limit of vitamin A.

10. The storage form of vitamin A in the liver is:

    (a) Retinyl palmitate.

    (b) Retinal.

    (c) Retinol.

    (d) Retinoic acid.

CHAPTER 4

Vitamin D
The Sunshine Vitamin

The main function of Vitamin D is regulating calcium and phosphorus to make bones strong. Vitamin D is an unusual vitamin. Vitamin D is necessary in the diet only for people who get too little sun to make their own vitamin D. For the billions of people who do get enough sun, it is not a vitamin because it is not needed in the diet. Vitamin D is also very hard to find in a natural diet as it occurs in only a few foods. Vitamin D is a fat-soluble vitamin. Vitamin D is also unusual because its most active form is one of the most powerful hormones in the human body.

Vitamin D was first named in 1922 by researchers who learned of a fat-soluble substance that played an important role in bone growth. Researchers went on to learn that ultraviolet light could activate vitamin D. By 1925 it was suspected that a cholesterol derivative in skin was activated by sunlight. In 1931 vitamin D_2 was synthesized, and the structure of vitamin D was established by 1936.

The Forms of Vitamin D

Vitamin D is found in four forms. *Cholecalciferol* is made by skin when the skin is exposed to direct sunlight containing the *B* form of ultraviolet radiation (*UVB*). Cholecalciferol is also called Vitamin D_3. It is the form used in many supplements and is sometimes used in food fortification.

> ## Forms of Vitamin D
>
> Vitamin D_3, cholecalciferol, is made in the skin or taken as a supplement.
> Calcidiol is the storage and circulating form of vitamin D.
> Calcitriol is the active form of vitamin D.
> Vitamin D_2, ergocalciferol, is made by irradiating fungi.

Calcidiol is made from cholecalciferol in the liver. Calcidiol is the form of vitamin D that is stored in the liver. It is a precursor to the hormone calcitriol. Blood tests for calcidiol can determine if there is enough vitamin D present or if there is a deficiency. The chemical name for calcidiol is 25-hydroxyvitamin D.

Calcitriol is made from calcidiol, primarily in the kidneys. It is the most potent steroid hormone derived from cholecalciferol. Calcitriol is the active form of vitamin D. It regulates calcium and possesses anti-cancer properties. The chemical name for calcitriol is 1,25-dihydroxyvitamin D.

Ergocalciferol, vitamin D_2, is made by irradiating ergosterol with ultraviolet light. Ergosterol is derived from the ergot fungus. It is a manmade product and not found in the human body unless it is taken as supplements or in fortified foods. Milk is often fortified with ergocalciferol in the amount of 100 IU per cup. Ergocalciferol has the same biological activity as cholecalciferol and can also be converted to calcidiol and calcitriol.

Sunlight and Vitamin D

Vitamin D_3, cholecalciferol, is made in the skin when a form of cholesterol (*7-dehydrocholesterol*) reacts with UVB ultraviolet light with wavelengths between 290 and 315 nanometers, as seen in Figure 4-1. Since UVB is absorbed

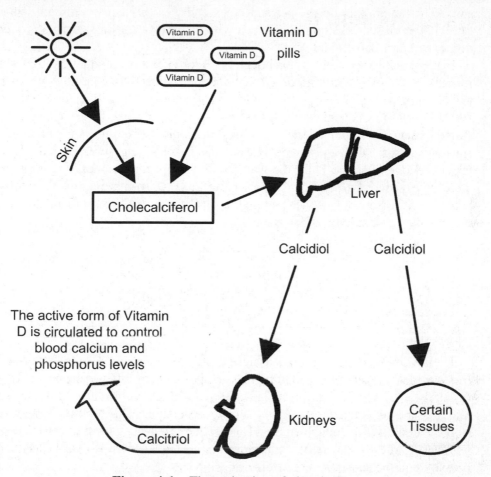

Figure 4-1 The activation of vitamin D.

by the atmosphere, more vitamin D is made when the sun is high. Fifteen minutes of summer sun in a bathing suit makes an average of 20,000 IU of vitamin D— 100 times the adequate daily intake. Since vitamin D is stored for long periods, this may be enough vitamin D to last for 100 days. UVB does not penetrate glass, so time in a closed car is not helpful.

Sunlight exposure provides most people with their entire vitamin D requirement. Young adults and children can make all of the vitamin D they need by spending just a few minutes in the sun three times a week. Older people have a slightly diminished capacity to synthesize vitamin D, so they need a little more sun. Also, many older people use sunscreen and wear protective clothing, which limits vitamin D production. Sunscreen with an SFP factor of 8 curtails production of vitamin D by 95 percent.

In higher latitudes, there is not enough UVB for vitamin D production in the skin for part of the year. Above 40 degrees north, approximately the latitude of Boston or San Francisco, there are four months without enough UVB to make vitamin D. Further north, in Canada, there are five months without much UVB. Vitamin D is stored in the liver for long periods of time. Even in higher latitudes, ten minutes of sun on the arms and face just three times weekly in the spring, summer, and fall will provide enough vitamin D for the whole year. Except for summer, it is important that this exposure be between 11:00 AM and 2:00 PM so that more UVB will be available. It requires twice the sunning time at 9:15 AM or 2:15 PM than at noon to provide the same amount of UVB exposure.

> Animals also produce vitamin D. Vitamin D is produced in fur and feathers. Animals lick or preen to absorb the vitamin.

Dark-skinned people living at higher latitudes may need extra vitamin D because their higher level of skin pigmentation may retard the absorption of UVB rays. Also, the weather in higher latitudes reduces daylight hours in winter and may require protective clothing which covers the skin. This is not a problem in tropical areas, where enough vitamin D is produced despite dark skin color. People with very dark skin may only produce one-sixth as much vitamin D as fair-skinned people do in the same amount of time.

Activation of Vitamin D

Vitamin D_3, whether taken as a supplement or made in the skin from sunlight, is biologically inactive. Cholecalciferol, vitamin D_3, is circulated to the liver through the bloodstream. In the liver cholecalciferol is *hydroxylated* (hydrogen and oxygen are added) to form calcidiol, the storage form of vitamin D. Calcidiol can remain in the liver for an extended time as the storage form of vitamin D. Calcidiol is also the major form of vitamin D found circulating in blood. Calcidiol is attached to a binding protein for transport through blood. With decreased sunlight on the skin, depleted

liver reserves, and not enough supplementation of vitamin D, calcidiol concentration in the blood decreases. This makes calcidiol a good indicator of vitamin D status.

> Calcitriol is one of the most potent steroid hormones in the human body.

To become active, this intermediary form of vitamin D, calcidiol, must be changed to its active form, calcitriol. This transformation to the active form normally takes place in the kidneys, but may also occur in some other tissues. It is only as calcitriol that vitamin D performs its physiological effects. Steroid hormones such as calcitriol are molecules in the body that are made from cholesterol. Steroid hormones act to turn certain genes on and off.

Calcitriol Regulates Genes That Make Proteins

Inside the nucleus of the cell, calcitriol assists in the transfer of genetic code information. One example of this is when calcitriol binds to a *transcription factor* that regulates the production of calcium transport proteins. These transport proteins assist calcium absorption in the intestines. Dozens of genes in cells throughout the body are regulated with calcitriol. Calcitriol works with the retinoic acid form of vitamin A and thyroid hormone to produce important proteins.

Calcium and phosphorus levels in the blood are maintained by the hormone calcitriol. The kidneys release just enough calcitriol into the blood to regulate blood levels of these minerals.

Calcitriol Inhibits the Proliferation of Cells

Some tissues in the body are able to make their own calcitriol from calcidiol. The calcitriol made in tissues outside the kidneys is not used to regulate calcium and phosphorus. The calcitriol made in tissues outside of the kidneys may play an anticancer role by slowing cell division.

How Vitamin D Controls Calcium

Blood calcium levels must be maintained within a narrow range for normal nervous system functioning and maintenance of bone density. Blood calcium levels are especially vital in childhood during bone growth. Vitamin D as calcitriol is an essential part of the regulation of blood calcium and phosphorus levels, as seen in Figure 4-2. Bone growth and regulation is also assisted by vitamin A, vitamin C, vitamin K, the hormone calcitonin, parathyroid hormone, and magnesium.

Calcium levels are sensed by the parathyroid glands. If blood levels of calcium fall too low, the parathyroid glands secrete parathyroid hormone. The parathyroid hormone stimulates production of an enzyme in the kidneys. This enzyme increases the transformation of calcidiol to calcitriol. Calcitriol is a potent hormone that increases blood calcium levels.

Increased blood levels of calcitriol cause increased absorption of calcium from food in the intestines. The kidneys also reduce losses of calcium in the urine in

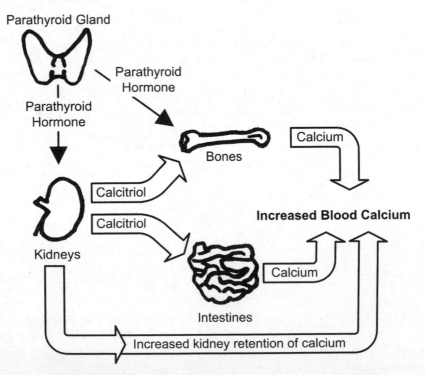

Figure 4-2   Vitamin D as calcitriol and blood calcium.

response to increased levels of calcitriol. In addition, calcium can be mobilized from bones if dietary levels of calcium are insufficient. Increased release of calcium from the bones requires parathyroid hormone in addition to calcitriol.

Vitamin D and Immunity

Calcitriol has a powerful ability to aid the functioning of the immune system. Immune system cells such as T cells and macrophage cells need calcitriol. Macrophages have the ability to make calcitriol from calcidiol. In autoimmune diseases an excess of macrophages may produce an excess amount of calcitriol.

Deficiency of Vitamin D

A deficiency of vitamin D makes it difficult for the body to keep enough calcium in the bloodstream. When there are insufficient levels of vitamin D, production of a protein that promotes uptake of calcium by cells lining the intestines slows. This allows calcium to pass through the intestines without being absorbed. The parathyroid glands sense the low blood levels of calcium and secrete more parathyroid hormone. The increasing levels of parathyroid hormone cause calcium to be released from the bones and retained in the kidneys.

When there is a severe deficiency of vitamin D in childhood, the bones fail to mineralize properly. Bones are most susceptible to low vitamin D levels when the bones are growing rapidly. With severe deficiency, the arms and legs become bowed—this is called *rickets*. Rickets can also result in delayed closure of the soft spots (fontanels) of the skull in infants. In very severe cases, low levels of blood calcium affect the nerves and seizures may result.

Deficiency Risk Factors

Breast-fed infants not exposed to sunlight.
The elderly have less ability to synthesize vitamin D
    in the skin.
Institutionalized people of all ages.
People with dark skin.
People who live in cold climates.

Vitamin D deficiency can also harm adults. Adult bones can slowly become deficient in calcium and phosphorus. This can result in soft bones, known as osteomalacia. Osteomalacia can cause a bent posture and bowed legs from chronic vitamin D deficiency.

Vitamin D deficiency can be a contributor to osteoporosis. Without enough vitamin D, the bones cannot properly mineralize. In addition, increased parathyroid hormone increases the loss of bone minerals. Lowered risk of osteoporotic fracture has been found when the daily intake of vitamin D was 600 to 700 IU as opposed to lower doses.

Sources of Vitamin D

The natural source of vitamin D is from sunshine. As you may recall, just a few minutes of sun on the face and arms a few times a week will provide all the vitamin D that you need, except in high latitudes during winter. It is best to get in the sun even in the winter, although reserves of stored vitamin D normally last through the winter even without sufficient sun to make more.

> Sunshine is the healthiest source of vitamin D.

There is no RDA for vitamin D. Instead, there is a daily Adequate Intake (AI). The AI for everyone under 50 years of age is 200 IU (5 mcg). Adults aged 51–70 have an AI of 400 IU (10 mcg) and for people over 70 the AI is 600 IU (15 mcg). These amounts are determined assuming that there is no vitamin D produced from sunlight. The AI is set to prevent rickets and osteomalacia.

Some experts recommend increased supplementation of vitamin D, while still staying well below the tolerable upper intake level (UL) of 2000 IU per day. These experts recommend an additional 200 IU on top of the AI, either from fortified foods, or from supplements. These higher levels may help to reduce the risk of osteoporosis and cancer.

Food sources of vitamin D are limited to a few fatty fish such as salmon, sardines, and mackerel. Some fish liver oils have high vitamin D content, but that is not the healthiest choice for vitamin D supplementation. Vitamin D is added to the feed of some laying hens to produce eggs with vitamin D. Some cereals, breads, and orange juices are fortified with vitamin D. Check the labels to see if they are fortified.

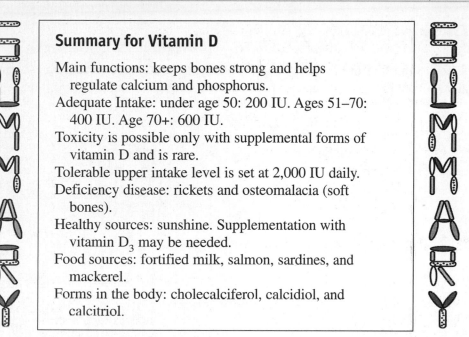

Summary for Vitamin D

Main functions: keeps bones strong and helps
   regulate calcium and phosphorus.
Adequate Intake: under age 50: 200 IU. Ages 51–70:
   400 IU. Age 70+: 600 IU.
Toxicity is possible only with supplemental forms of
   vitamin D and is rare.
Tolerable upper intake level is set at 2,000 IU daily.
Deficiency disease: rickets and osteomalacia (soft
   bones).
Healthy sources: sunshine. Supplementation with
   vitamin D_3 may be needed.
Food sources: fortified milk, salmon, sardines, and
   mackerel.
Forms in the body: cholecalciferol, calcidiol, and
   calcitriol.

Milk is fortified with the addition of 400 IU of vitamin D_2 to each quart in the United States. Other dairy products are not normally fortified. Margarine may be fortified with Vitamin D.

Toxicity of Vitamin D

It is important to note that vitamin D created from sunlight on skin is not known to result in toxic levels. Levels of vitamin D produced from sunlight are self-adjusting.

Vitamin D toxicity is known as *hypervitaminosis D*. Toxic levels of vitamin D can cause abnormally high blood calcium levels. This can result in bone loss and kidney stones. Long-term overconsumption of vitamin D can cause calcification of organs such as the heart, blood vessels, and the kidneys.

Vitamin D toxicity is unlikely in healthy adults with supplement levels lower than 10,000 IU/day. The Food and Nutrition Board has established a very conservative tolerable upper intake level (UL) of 2,000 IU/day (50 mcg/day) for children and adults. The UL for infants up to one year of age is 1000 IU.

To summarize, we need vitamin D to keep our bones strong. Vitamin D may play an anti-cancer role by slowing cell division. Sunshine is the natural source of vitamin D.

Quiz

Refer to the text in this chapter if necessary. A good score is at least 8 correct answers out of these 10 questions. The answers are listed in the back of this book.

1. The form of vitamin D found in fortified food is:
   (a) Cholecalciferol.
   (b) Calcidiol.
   (c) Calcitriol.
   (d) Ergocalciferol.

2. The most active form of vitamin D is:
   (a) Cholecalciferol.
   (b) Calcidiol.
   (c) Calcitriol.
   (d) Ergocalciferol.

3. The natural source of vitamin D is:
   (a) Sunshine.
   (b) Spinach.
   (c) Apples.
   (d) Vitamin D supplements.

4. Vitamin D is metabolized and activated in:
   (a) The liver.
   (b) The kidneys.
   (c) The skin.
   (d) All of the above.

5. Blood calcium can come from:
   (a) The bones.
   (b) The intestines.
   (c) The kidneys.
   (d) All of the above.

6. The deficiency disease for vitamin D is:
   (a) Pellagra.
   (b) Rickets.
   (c) Scurvy.
   (d) Night blindness.

7. The tolerable upper intake level (UL) for vitamin D is:
   (a) 200 IU.
   (b) 400 IU.
   (c) 1000 IU.
   (d) 2000 IU.

8. Older people need:
   (a) More vitamin D.
   (b) Less vitamin D.
   (c) The same amount of vitamin D.
   (d) No vitamin D.

9. Vitamin D helps regulate:
   (a) Iron.
   (b) Calcium.
   (c) Potassium.
   (d) Magnesium.

10. The gland (glands) that works (work) with vitamin D to regulate blood calcium is (are):
    (a) Thyroid gland.
    (b) Parathyroid glands.
    (c) Adrenal glands.
    (d) Thymus gland.

CHAPTER 5

Vitamin E
The Fat Antioxidant

The main form of vitamin E is *alpha-tocopherol*. *Tocopherols* were discovered at the University of California at Berkeley in 1922, where they were found to be essential to maintain fertility. Pure alpha-tocopherol was first isolated from wheat-germ oil in 1936. The word tocopherol means "to bear offspring" and derives from the Greek root *phero*, which means "to bring forth," and the Greek root *tos*, which means "childbirth." Tocopherols are a family of eight fat-soluble alcohols. The final "ol" in the name tocopherol indicates that it is an alcohol.

The Forms of Vitamin E

The tocopherols are divided into four types: alpha-tocopherol, beta-tocopherol, gamma-tocopherol, and delta-tocopherol, as shown in Figure 5-1. Also in the vitamin E family are the very similar *tocotrienols*. These tocotrienols are also divided into four groups: alpha-, beta-, gamma-, and delta-tocotrienol. All tocopherols and

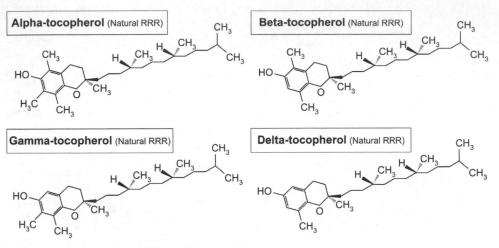

Figure 5-1   Tocopherol structures.

all tocotrienols have the ability to donate a hydrogen atom to neutralize free radicals. They also have a side chain with the ability to penetrate biological membranes and fats.

ALPHA-TOCOPHEROL

Alpha-tocopherol is the form of vitamin E found in the largest quantities in blood and tissue inside the human body. Alpha-tocopherol is a powerful antioxidant that protects membranes and *low density lipoproteins* (LDL) from free radical damage. Alpha-tocopherol is the only form of vitamin E that the human body actively works to keep in the bloodstream. The RDAs for vitamin E are based upon levels of alpha-tocopherol.

GAMMA-TOCOPHEROL

Gamma-tocopherol is the type of vitamin E found in the greatest abundance in most American diets. Blood levels of gamma-tocopherol are typically only 10 percent of the blood levels of alpha-tocopherol, despite the larger dietary intake of gamma-tocopherol. On the other hand, certain tissues do have significant concentrations of gamma-tocopherol. Gamma-tocopherol makes up about 40 percent of the tocopherols in muscles and about a third of the tocopherols in veins. More research is needed to confirm if gamma-tocopherol is effective in reducing the risk of prostate cancer. Gamma-tocopherol may be more potent than alpha-tocopherol as an anticoagulant.

MIXED TOCOPHEROLS

There is general agreement that increased levels of vitamin E in the diet reduce the risk of cardiovascular disease. However, studies on the cardiovascular protective effects of supplementary vitamin E given as synthetic alpha-tocopherol have had inconsistent results. Research continues, but there are indications that the other forms of tocopherols, collectively called *mixed tocopherols*, are also important as antioxidants and anticoagulants.

TOCOTRIENOLS

Tocotrienols are much more potent antioxidants than tocopherols. They are very similar to tocopherols, as seen in Figure 5-2. Unfortunately, they are poorly assim-ilated and rapidly eliminated from the body. The tocotrienols are well-absorbed by skin, so they are good choices for antioxidants in lotions. Vitamin E is sometimes used to reduce scarring from healing injuries after the skin has closed. The tocotrienol form of vitamin E may be the correct choice for this type of topical application. Palm oil is a rich source of tocotrienols.

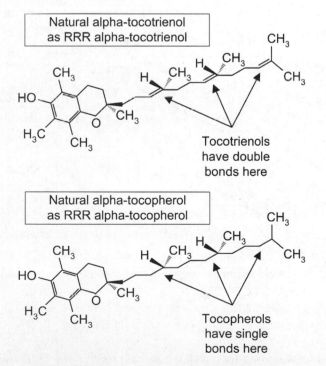

Figure 5-2   Tocopherols and tocotrienols.

Antioxidant Activity

The main role of vitamin E is as an antioxidant to neutralize free radicals in cell membranes, in mitochondrial membranes, and in LDL. The fats in cell membranes are susceptible to oxidation by free radicals. The interior of the cell membrane is inaccessible to water-soluble antioxidants. The fat-soluble tocopherols are perfectly suited to protecting cell membranes from free radicals. The fat-soluble tail of the tocopherol can reach in and neutralize the free radicals, as seen in Figure 5-3. Vitamin E also protects polyunsaturated fatty acids and vitamin A from free radical damage.

Vitamin E neutralizes free radicals by donating a hydrogen atom from the hydroxyl group (HO) on the hexagonal head of vitamin E, as seen in Figure 5-4. Donation of hydrogen to a free radical is easiest for the alpha-tocopherol and

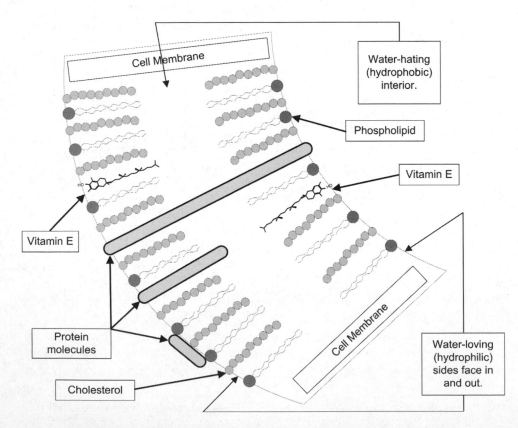

Figure 5-3   Tocopherol can reach deep inside the cell membrane.

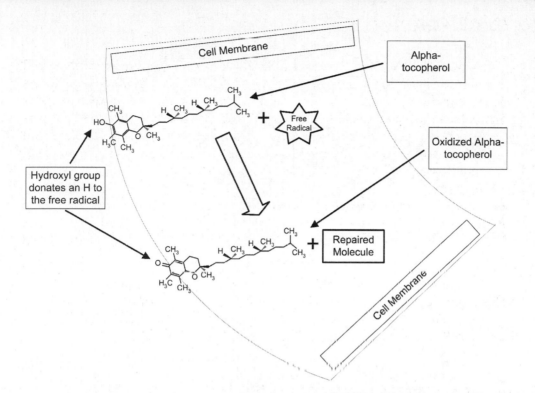

Figure 5-4   Free radical neutralization in a cell membrane with tocopherol.

gamma-tocopherol forms and slightly harder for the beta-tocopherol and delta-tocopherol forms.

In cell membranes, the hexagonal head of vitamin E stays near the surface of the membrane. The tail of vitamin E is deeply implanted into the cell membrane. When the tail of vitamin E is oxidized by a free radical, it moves to the surface where the head of the vitamin E can donate hydrogen. In this way, vitamin E pulls free radicals out of fatty membranes for neutralization. This vitamin E cannot perform its antioxidant function a second time until it is regenerated.

The ability of vitamin E to perform as an antioxidant can be restored by the ascorbate form of vitamin C or by coenzyme Q. After restoring the antioxidant activity of vitamin E, the vitamin C then needs glutathione to restore its own antioxidant activity, as seen in Figure 5-5. In reactivating these antioxidants, both niacin and lipoic acid may play a role. Vitamin E has the potential to act as a free radical rather than as an antioxidant when co-antioxidants such as vitamin C are not available.

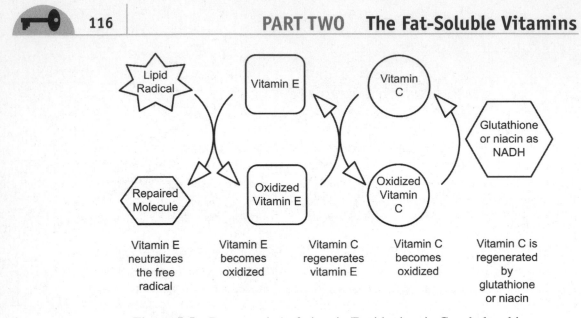

Figure 5-5   Regeneration of vitamin E with vitamin C and glutathione or niacin.

Cholesterol and Vitamin E

Proteins are attached to fats to facilitate their transport though the bloodstream. One example is a waxy fat, cholesterol, which is transported in the blood attached to proteins in the form of lipoproteins. Low density lipoproteins (LDL) transport cholesterol away from the liver, while the *high density lipoproteins* (HDL) transport cholesterol back to the liver for storage or for elimination in the bile. There is a close association between elevated levels of LDL cholesterol and coronary heart disease, which is a leading cause of death all over the world.

Low density lipoproteins are assembled in the liver. They are not known to contribute to heart disease unless they are oxidized by free radicals. The creation of atherosclerotic plaque and its progression to heart disease is only partially understood. When LDL levels in the blood are elevated, the oxidation of LDL may lead to the accumulation of fatty slabs on the inside of the arterial walls. This is the leading theory of the development of atherosclerotic plaque. The elevated levels of oxidized LDL stimulate macrophages to attack the LDL using free radicals. This causes more oxidized LDL. The macrophages become bloated foam cells and these foam cells attach to the inside of the artery. This can lead to blockage of the coronary arteries and a heart attack.

> Vitamin E is built into LDL for protection from oxidation that can lead to clogged arteries.

Four published, large-scale, randomized, double-blind clinical intervention studies tested the effectiveness of supplemental vitamin E to reduce heart attacks. Three of these studies used 50 mg of vitamin E in the synthetic form and found little correlation with heart attacks. However, the study using 268 mg to 567 mg of vitamin E in the natural alpha-tocopherol form found a strong correlation with reduced heart attacks.

Vitamin E protects LDL from oxidation. In the liver, a protein called *alpha-tocopherol transfer protein* incorporates alpha-tocopherol into LDL during the assembly of LDL. The alpha-tocopherol is there to protect the LDL from free radicals. Although alpha-tocopherol is preferentially loaded into LDL, some of the other tocopherols can also be incorporated into LDL. Half of the synthetic forms of vitamin E cannot be loaded into LDL.

Vitamin E and Blood Circulation

Alpha-tocopherol decreases clumping of blood (platelet aggregation). This eases the passage of blood through capillaries and narrowed blood vessels. Beta-tocopherol has not been found to decrease platelet aggregation.

Alpha-tocopherol has a regulatory effect on the tone of the muscles of the circulatory system. As the heart pumps, blood vessels must expand and contract. Alpha-tocopherol supplementation has been correlated with maintaining normal arterial wall flexibility.

Deficiency of Vitamin E

True vitamin E deficiency is rare in the United States and Canada. However, the consumption of less than optimal amounts of vitamin E is common. About one-third of adults were found to have blood levels of vitamin E at levels so low as to increase their risk of cardiovascular diseases. These low levels of vitamin E are normally due to low dietary intakes. People who have difficulties absorbing fat are even more likely to have low levels of vitamin E. For example, people with cystic fibrosis may have low levels of vitamin E because of poor absorption of fats.

> About one-third of adults were found to have
> blood levels of vitamin E at levels so low as to
> increase their risk of cardiovascular diseases.

With severe deficiency of vitamin E, red blood cell membranes can rupture. This may be because of oxidation of the cell membranes due to a lack of vitamin E. Severe, prolonged vitamin E deficiency can cause neurological problems including impaired vision and poor muscle coordination.

Food Sources of Vitamin E

Average dietary intake of vitamin E in the United States is not enough to meet the RDAs. The average daily intake is about 14 IU (9 mg) for men and 9 IU (6 mg) for women. This is well below the RDA of 22.5 IU (15 mg) for adults. It is not easy to get sufficient vitamin E in a typical American diet without excessive fat intake.

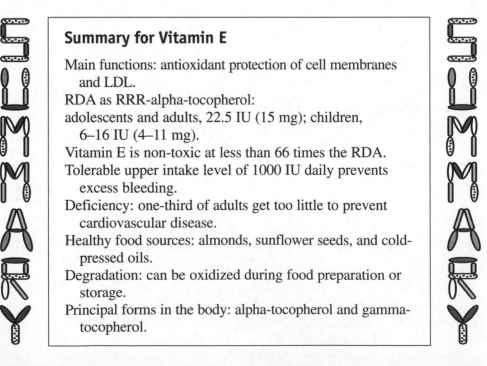

Summary for Vitamin E

Main functions: antioxidant protection of cell membranes
    and LDL.
RDA as RRR-alpha-tocopherol:
adolescents and adults, 22.5 IU (15 mg); children,
    6–16 IU (4–11 mg).
Vitamin E is non-toxic at less than 66 times the RDA.
Tolerable upper intake level of 1000 IU daily prevents
    excess bleeding.
Deficiency: one-third of adults get too little to prevent
    cardiovascular disease.
Healthy food sources: almonds, sunflower seeds, and cold-
    pressed oils.
Degradation: can be oxidized during food preparation or
    storage.
Principal forms in the body: alpha-tocopherol and gamma-
    tocopherol.

Vitamin E is found in fatty foods such as vegetable oils, nuts, and seeds; please refer to Graph 5-1. Smaller amounts of vitamin E are also found in whole grains,

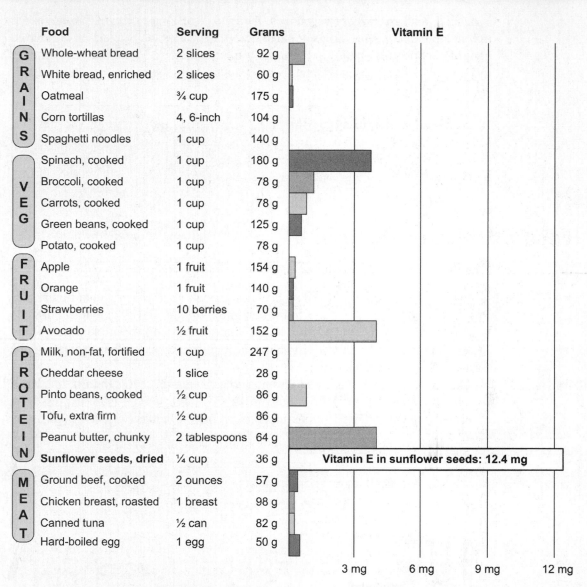

Food	Serving	Grams	Vitamin E
GRAINS			
Whole-wheat bread	2 slices	92 g	
White bread, enriched	2 slices	60 g	
Oatmeal	¾ cup	175 g	
Corn tortillas	4, 6-inch	104 g	
Spaghetti noodles	1 cup	140 g	
VEG			
Spinach, cooked	1 cup	180 g	
Broccoli, cooked	1 cup	78 g	
Carrots, cooked	1 cup	78 g	
Green beans, cooked	1 cup	125 g	
Potato, cooked	1 cup	78 g	
FRUIT			
Apple	1 fruit	154 g	
Orange	1 fruit	140 g	
Strawberries	10 berries	70 g	
Avocado	½ fruit	152 g	
PROTEIN			
Milk, non-fat, fortified	1 cup	247 g	
Cheddar cheese	1 slice	28 g	
Pinto beans, cooked	½ cup	86 g	
Tofu, extra firm	½ cup	86 g	
Peanut butter, chunky	2 tablespoons	64 g	
Sunflower seeds, dried	¼ cup	36 g	Vitamin E in sunflower seeds: 12.4 mg
MEAT			
Ground beef, cooked	2 ounces	57 g	
Chicken breast, roasted	1 breast	98 g	
Canned tuna	½ can	82 g	
Hard-boiled egg	1 egg	50 g	

3 mg   6 mg   9 mg   12 mg

Graph 5-1   Vitamin E in some common foods.

avocados, and green leafy vegetables. Almonds, safflower oil, and hazelnuts are rich in alpha-tocopherol. Soybean oil, corn oil, avocados, and canola oil are rich in gamma-tocopherol. These foods naturally have a mixture of the various forms of tocopherols and tocotrienols, but always contain only the natural isomer of each.

Foods Rich in Vitamin E (mg per 100 grams)			
Sunflower oil	59	Soybean oil	18
Sunflower seeds	50	Hazelnuts	15
Safflower oil	43	Olive oil	12
Almonds	26	Spinach, raw	2
Corn oil	21	Avocados	1
Canola oil	21	Whole wheat	1

Vitamin E is easily destroyed by heat and oxidation, as illustrated in Figure 5-6. Processed foods are a poor choice for obtaining vitamin E. For example, enriched white flour has only two percent of the vitamin E found in whole wheat flour. Deep frying not only destroys vitamin E, but causes extra free radicals to be formed in the heated oil.

The RDA for vitamin E for adults and adolescents is 22.5 IU (15 mg). The RDA for children ranges from 6 IU (4 mg) to 16 IU (10 mg). RDAs are expressed in amounts of the natural RRR-alpha-tocopherol form, which is twice as potent as the

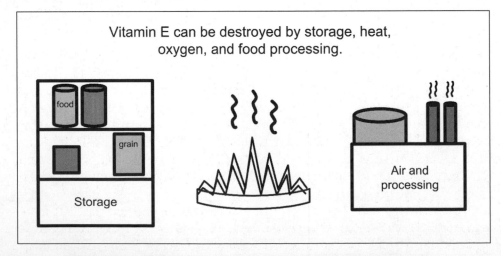

Figure 5-6   Vitamin E is easily destroyed during food processing.

Table 5-1   RDAs for vitamin E for all ages.

RDAs for Vitamin E	Age	Males mg/day	Females mg/day
Infants	0–6 months	4 mg (6 IU)	4 mg (6 IU)
Infants	7–12 months	5 mg (7.5 IU)	5 mg (7.5 IU)
Children	1–3 years	6 mg (9 IU)	6 mg (9 IU)
Children	4–8 years	7 mg (10.5 IU)	7 mg (10.5 IU)
Children	9–13 years	11 mg (16.5 IU)	11 mg (16.5 IU)
Adolescents	14–18 years	15 mg (22.5 IU)	15 mg (22.5 IU)
Adults	19 years and older	15 mg (22.5 IU)	15 mg (22.5 IU)
Pregnancy	all ages	—	15 mg (22.5 IU)
Breastfeeding	all ages	—	19 mg (28.5 IU)

synthetic forms. Twice as many milligrams of the synthetic forms are needed to provide the same activity as the natural form.

The following RDAs are based on the prevention of deficiency symptoms such as the breakdown of red blood cells. The RDAs are not based on the prevention of chronic diseases or on the promotion of optimum health. Please refer to Table 5-1 for RDAs for all ages.

Vitamin E Supplements

Vitamin E supplements must be taken with food for proper absorption. Natural vitamin E in supplements is designated by a *d* as in d-alpha-tocopherol. Synthetic vitamin E is designated with a *dl* as in dl-alpha-tocopherol. The newer and more accurate designations differentiate between *R* for natural and *S* for synthetic. As seen in Figure 5-7, the synthetic forms are different from the natural forms. There are eight possible isomers: RRR (natural), SRR, SSR, SRS, SSS, RSR, RRS, and RSS. Synthetic vitamin E is a mix of eight isomers, of which only one is natural. Half of the synthetic forms (SRR, SSR, SRS, and SSS) do not function as vitamin E in the body at all.

The correct, current way to describe natural alpha-tocopherol is RRR-alpha-tocopherol. When vitamin E is synthesized, all eight isomers are made and the result is called "*racemic*," as in *all-rac-alpha-tocopherol*. This synthetic mix is the most common form used in medical studies.

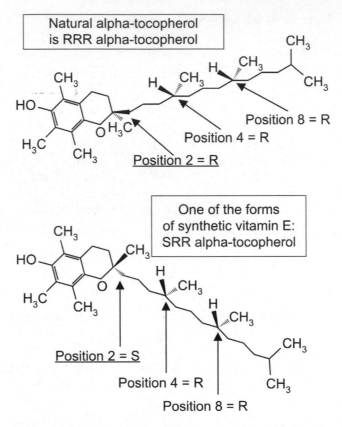

Figure 5-7   Natural and synthetic alpha-tocopherol.

VITAMIN E POTENCY

There are two ways of stating the potency of vitamin E, in International Units (IU) and in milligrams (mg). The acetate form of synthetic alpha-tocopherol was arbitrarily chosen to have one IU per mg. Natural RRR-alpha-tocopherol has a biological activity of at least 1.49 IU per mg. Continuing research is indicating that the biological activity of natural vitamin E (RRR-alpha-tocopherol) may be 2 IU per mg or more. Natural vitamin E has more activity because it is not diluted with synthetic isomers. In synthetic all-rac-alpha-tocopherol, only half of the isomers are usable by the body, so it is less potent. Only one of the eight isomers in synthetic alpha-tocopherol is found in natural food.

VITAMIN E IN SUPPLEMENTS

There are two common forms of synthetic alpha-tocopherol used in supplements: the *acetate ester* form and the *succinate ester* form. Both forms are broken apart into alpha-tocopherol in the intestines. Acetate ester and succinate ester are more stable in storage because they are not antioxidants until they are de-esterified in the intestines.

Both of these synthetic forms contain one natural form of alpha-tocopherol and seven isomers of alpha-tocopherol that are not found in nature. These synthetic supplements also differ from natural vitamin E because they are missing all other tocopherols and tocotrienols. The inconsistent results in medical studies with these synthetic supplements may be partially due to these differences between vitamin E found in food and vitamin E found in synthetic supplements.

Toxicity of Vitamin E

Vitamin E is not toxic, but, in very large doses, it may impair blood clotting. The tolerable upper intake level (UL) for natural alpha-tocopherol is 1000 mg per day, about sixty-six times the RDA. The UL for synthetic alpha-tocopherol is also 1000 mg per day. This upper level is conservative, as few side effects have been noted below 3000 IU (2000 mg). For optimal health and disease resistance, 400 to 800 IU of natural vitamin E is often taken. It is recommended that supplementation of vitamin E over 400 IU be increased gradually so that the body can get used to the blood-thinning effects, especially in cases of high blood pressure.

Some surgeons recommend the discontinuation of vitamin E supplementation before surgery to decrease any risk of excess bleeding. Premature infants are very sensitive to alpha-tocopherol supplements. The UL for infants under one year has not been established, so supplementation of vitamin E in infants should only be attempted under the close supervision of a pediatrician.

People taking anticoagulant drugs should not take vitamin E supplements. People deficient in vitamin K should also avoid vitamin E supplementation to avoid the possibility of excess bleeding.

To summarize, as the principal fat-soluble antioxidant, vitamin E is needed by every cell and artery. Special care must be taken to obtain enough vitamin E from food.

Quiz

Refer to the text in this chapter if necessary. A good score is at least 8 correct answers out of these 10 questions. The answers are listed in the back of this book.

1. Which is NOT a form of vitamin E?
   (a) Alpha-tocopherol.
   (b) Alpha-tocotrienol.
   (c) Omega-tocopherol.
   (d) Gamma-tocopherol.

2. Vitamin E neutralizes free radicals in:
   (a) Bones.
   (b) Membranes.
   (c) Hair.
   (d) Gallbladder.

3. Vitamin E can be reactivated by:
   (a) Vitamin C.
   (b) Vitamin A.
   (c) B vitamins.
   (d) Vitamin D.

4. Vitamin E helps circulation by:
   (a) Protecting LDL from oxidation.
   (b) Increasing capillary flexibility.
   (c) Decreasing blood coagulation.
   (d) All of the above.

5. Vitamin E supplements, to be effective, should be:
   (a) Taken between meals.
   (b) Taken with meals.
   (c) Taken one half-hour before meals.
   (d) Taken before bedtime.

6. Good food sources of vitamin E are:
   (a) Cold-pressed oils.
   (b) Fruit.
   (c) Enriched flour products.
   (d) French fries.

7. Natural vitamin E is:
   (a) SRR-alpha-tocopherol.
   (b) All-rac-alpha-tocopherol.
   (c) RRR-alpha-tocopherol.
   (d) Dl-alpha-tocopherol.

8. Vitamin E supplementation at over 66 times the RDA:
   (a) Can cause yellowed skin.
   (b) Can cause diabetes.
   (c) Can cause rickets.
   (d) Can cause excessive bleeding.

9. Tocotrienols:
   (a) Are a form of vitamin E.
   (b) Are potent antioxidants.
   (c) Are absorbed through skin.
   (d) All of the above.

10. Vitamin E is:
    (a) Fat-soluble.
    (b) Water-soluble.
    (c) Mineral-soluble.
    (d) Not soluble.

CHAPTER 6

Vitamin K
The Green Leafy Vitamin

Vitamin K is a fat-soluble vitamin needed for blood *coagulation*. Coagulation refers to the process of blood clotting. Vitamin K was discovered by a Danish scientist, Henrik Dam, in the late 1920s. He discovered a factor that was causing excessive bleeding and was missing from some diets. He published his work in a German journal and called the new coagulation vitamin *Koagulationsvitamin*. The initial letter in this word is how vitamin K got its name.

The Forms of Vitamin K

Plants synthesize *phylloquinone*, which is known as *vitamin K_1*. Bacteria can also synthesize vitamin K in several forms. The vitamin K made by bacteria is known as *menaquinone*, or *vitamin K_2*. There are many forms of menaquinone made by bacteria, each with different numbers of side chains. The different forms of vitamin K have the same action. However, they may vary in absorption and tissue distribution. Intestinal absorption is influenced by the food source of the vitamin. The varying side chains on vitamin K affect the fat solubility, and this affects tissue distribution.

> **The Forms of Vitamin K**
>
> Phylloquinone   Vitamin K$_1$
> Menaquinone   Vitamin K$_2$
> Menadione   Vitamin K$_3$ (no longer used)

Certain proteins are able to bind calcium with the assistance of vitamin K. *Glutamic acid* is one of the amino acids used to make certain proteins. Vitamin K converts the glutamic acid residues in certain proteins to a form of glutamic acid that can bind calcium. Vitamin K is a coenzyme needed by an enzyme that converts glutamic acid residues into g*amma-carboxyglutamic acid*. Once vitamin K has converted the glutamic acid, the protein is then able to bind calcium. There are more than one dozen proteins affected by vitamin K. These proteins are vital for blood coagulation and bone strength.

Vitamin K and Blood Clotting

Blood must flow freely and not clot unless there is a break in a blood vessel. Clotting is normally triggered by a rough edge in a blood vessel, such as a cut artery. The liver synthesizes several proteins important in blood clotting including *prothrombin* and *fibrinogen*. These two proteins circulate in blood. One of the first steps in blood clotting is triggered by the formation of *thrombin* from prothrombin, as seen in Figure 6-1. Thrombin then accelerates the conversion of fibrinogen to *fibrin*, which consists of fine threads that tangle together to form a blood clot. Vitamin K is necessary in order for the liver to synthesize prothrombin. Vitamin K converts the glutamic acid that is in prothrombin, enabling it to bind to calcium. Without adequate vitamin K, prothrombin production slows, leading to a bleeding tendency. There are several other proteins involved in blood clotting that are activated by vitamin K.

Balance and control are important parts of blood clotting. In addition to speeding blood clotting, vitamin K also assists in creating proteins that slow blood clotting. Vitamin K is needed to synthesize two proteins, *protein C* and *protein S*,

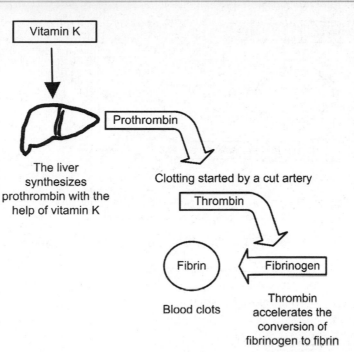

Figure 6-1   The role of vitamin K in blood clotting.

which are important inhibitors of coagulation. Protein S is also synthesized by the blood vessel walls where it has a role as a coagulation inhibitor.

Vitamin K and Bone Mineralization

Bones are formed by bone-forming cells called *osteoblasts*. The osteoblast cells synthesize *osteocalcin*, under the direction of the active form of vitamin D, calcitriol. Vitamin K is needed to enable osteocalcin to bind minerals to bones, as diagrammed in Figure 6-2. Vitamin K is used as a coenzyme to convert three glutamic acid residues in osteocalcin that enable bone mineralization.

Higher vitamin K levels may be protective against osteoporosis and age-related fracture. Several studies have found a correlation between higher vitamin K levels and lowered risk of hip fracture. However, since leafy green vegetables are the primary source of vitamin K, this protective effect could be from other nutrients, such as the calcium or magnesium in green leafy vegetables.

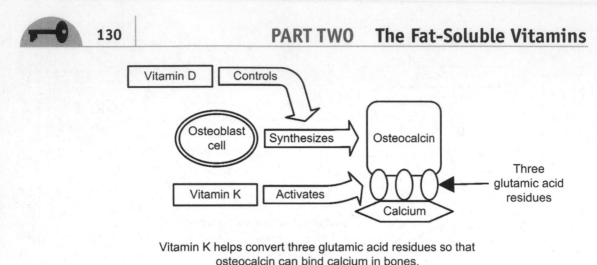

Vitamin K helps convert three glutamic acid residues so that
osteocalcin can bind calcium in bones.

Figure 6-2   Vitamin K and bone mineralization.

Deficiency of Vitamin K

Healthy adults rarely have a deficiency of vitamin K. Adult deficiency is normally not seen unless there is liver disease or if anticoagulant drugs are being used. Vitamin K is abundant in food. In addition, the bacteria in the large intestine synthesize vitamin K_2, menaquinone. It is not clear how much of this vitamin K_2 is absorbed and used. As much as half of the daily requirement may be produced by these bacteria. One more reason that deficiency is rare is that vitamin K can be used over and over again in a conservation cycle.

> **Why Adult Vitamin K Deficiency is Rare**
>
> Vitamin K is common in food.
> Bacteria make vitamin K in the large intestine.
> Vitamin K is reused in a conservation cycle.

Vitamin K deficiency can be determined by laboratory tests that measure clotting time. Blood clotting may be slower when there is a deficiency of vitamin K. Symptoms of vitamin K deficiency include nosebleeds, bleeding gums, prolonged bleeding from cuts, or blood in the urine or stool.

Infant Vitamin K Deficiency

Infant vitamin K deficiency can cause intracranial hemorrhage (bleeding inside the skull) and can be life-threatening. As many as one in five thousand infants may develop vitamin K deficiency bleeding unless supplemental vitamin K is given. There are three reasons why infants may have low levels of vitamin K. The first reason is that breast milk is normally low in vitamin K. Newborn infants are estimated to receive about ten percent of their recommended daily intake when breastfeeding. Normal doses of supplementary vitamin K given to the nursing mother do not seem to elevate breast milk levels of vitamin K. The second reason is that the newborn's intestines may not contain the bacteria that synthesize menaquinone (vitamin K_2). Finally, infants may not have fully developed their vitamin K conservation cycle.

> ### Why Newborn Infants May be Deficient in Vitamin K
>
> Vitamin K levels in breast milk are low.
> Bacteria to make vitamin K in the large intestine may not yet
>   be ready.
> The vitamin K conservation cycle may not be developed.

Because of these reasons, newborn infants are routinely given an injection of 1000 mcg of vitamin K_1 (phylloquinone). Some infants receive one to three oral doses of vitamin K instead, which is almost as effective as the single injection. A single injection results in extremely high levels of vitamin K in the blood. An infant's blood level of vitamin K may go up to 9000 times the normal adult level, tapering off to 100 times the normal adult level after four days. Lower doses of vitamin K are recommended for premature infants.

For mothers who refuse vitamin K injections for their infant, several oral doses of vitamin K are an alternative. If neither is chosen, dietary changes may reduce the possibility of vitamin K deficiency bleeding. While extra vitamin K intake during pregnancy does not increase vitamin K in the unborn child, large amounts of vitamin K intake during breastfeeding can increase the infant's blood vitamin K levels. The vitamin K deficiency bleeding problems normally occur one to seven weeks after birth. Mothers would need to eat a cup of broccoli or other vitamin K-rich food and a large green salad coupled with extra supplementation to boost the infant's vitamin K levels. Careful monitoring of bleeding tendencies in newborns would also be needed by mothers who refuse or who are unable to use either vitamin K injections or oral dosing.

Food Sources of Vitamin K

Green leafy vegetables are the primary source of dietary vitamin K; please refer to Graph 6-1. Certain oils contribute a little vitamin K to the diet, including oils of olive, canola, and soybean. Kale and collards are excellent sources, with just one cup providing about ten times the dietary reference intake (DRI). Spinach and beet greens are also rich sources of vitamin K. These vegetables are also high in other important micronutrients.

Summary for Vitamin K

Main functions: prevents excess bleeding and assists in bone mineralization.

Daily Recommended Intake: men, 120 mcg; women, 90 mcg.

Vitamin K is non-toxic. No tolerable upper level has been set.

Deficiency: rare in adults, may occur in newborn infants.

Food sources: green leafy vegetables are the best source.

Principal forms in the body: phylloquinone (vitamin K_1) and menaquinone (vitamin K_2)

Vitamin K is absorbed from the intestines with the help of bile salts. About 80 percent of dietary phylloquinone (K_1) is absorbed. The intestinal mucosa prepares the vitamin K for transport by the lymph system. The body stores very

The Richest Sources of Vitamin K in Micrograms (mcg)

Kale	1 cup	**1147**
Collards	1 cup	**1060**
Spinach	1 cup	**1027**
Beet greens	1 cup	**697**
Brussels sprouts	1 cup	**300**
Lettuce	1 head	**167**
Parsley	10 sprigs	**164**
Cabbage	1 cup	**163**
Soybean oil	1 Tablespoon	**26**

	Food	Serving	Grams	Vitamin K
GRAINS	Whole-wheat bread	2 slices	92 g	
	White bread, enriched	2 slices	60 g	
	Oatmeal	¾ cup	175 g	
	Corn tortillas	4, 6-inch	104 g	
	Spaghetti noodles	1 cup	140 g	
VEG	**Spinach, cooked**	1 cup	180 g	Vitamin K in spinach: 1027 mg
	Broccoli, cooked	1 cup	78 g	
	Carrots, cooked	1 cup	78 g	
	Green beans, cooked	1 cup	125 g	
	Potato, cooked	1 cup	78 g	
FRUIT	Apple	1 fruit	154 g	
	Orange	1 fruit	140 g	
	Strawberries	10 berries	70 g	
	Avocado	½ fruit	152 g	
PROTEIN	Milk, non-fat, fortified	1 cup	247 g	
	Cheddar cheese	1 slice	28 g	
	Pinto beans, cooked	½ cup	86 g	
	Tofu, extra firm	½ cup	86 g	
	Peanut butter, chunky	2 tablespoons	64 g	
	Sunflower seeds, dried	¼ cup	36 g	
MEAT	Ground beef, cooked	2 ounces	57 g	
	Chicken breast, roasted	1 breast	98 g	
	Canned tuna	½ can	82 g	
	Hard-boiled egg	1 egg	50 g	

0.125 mg   0.25 mg   0.375 mg   1000 mg

Graph 6-1   Vitamin K in some common foods.

Table 6-1   Dietary Reference Intakes for vitamin K for all ages.

Reference Intakes for Vitamin K	Age	Males mcg/day	Females mcg/day
Infants	0–6 months	2.0	2.0
Infants	7–12 months	2.5	2.5
Children	1–3 years	30	30
Children	4–8 years	55	55
Children	9–13 years	60	60
Adolescents	14–18 years	75	75
Adults	19 years and older	120	90
Pregnancy	18 years and younger	—	75
Pregnancy	19 years and older	—	90
Breastfeeding	18 years and younger	—	75
Breastfeeding	19 years and older	—	90

little vitamin K and regular dietary intake is necessary. Most of the vitamin K is stored in the liver. Vitamin K is constantly lost via the bile and also in urine. The conservation cycle allows some vitamin K to be reused.

Supplemental vitamin K is in the form of phylloquinone, vitamin K_1. Normal doses range from 10 mcg to 120 mcg. The best way to supplement vitamin K is to eat abundant amounts of green leafy vegetables. Please refer to Table 6-1 for reference intakes of daily amounts of vitamin K. One of the forms of vitamin K_2 is called *menatetrenone* (MK-4), and is being investigated for possible benefit in treating osteoporosis.

Toxicity of Vitamin K

No upper limit has been set for vitamin K. There is no known toxicity with high doses of vitamin K_1 or vitamin K_2. One form of vitamin K, menadione (vitamin K_3), is no longer used for vitamin K deficiency because it may interfere with glutathione, an important antioxidant.

To summarize, vitamin K is needed to prevent excess bleeding in case of injury. Vitamin K is also important for strong bones. Vitamin K is easy to find in green, leafy vegetables.

Quiz

Refer to the text in this chapter if necessary. A good score is at least 8 correct answers out of these 10 questions. The answers are listed in the back of this book.

1. The main function of vitamin K is:
    (a) Blood thinning.
    (b) Blood coagulation.
    (c) Antioxidant.
    (d) Energy metabolism.

2. The form of vitamin K found in green vegetables is:
    (a) Menaquinone.
    (b) Menadione.
    (c) Menatetrenone.
    (d) Phylloquinone.

3. The form of vitamin K produced by bacteria in the large intestine is:
    (a) Menaquinone.
    (b) Menadione.
    (c) Menatetrenone.
    (d) Phylloquinone.

4. Certain proteins are able to bind calcium after vitamin K has altered the:
    (a) Lysine.
    (b) Arginine.
    (c) Glycine.
    (d) Glutamic acid.

5. Without enough vitamin K, the body cannot produce enough:
    (a) Fibrin.
    (b) Thrombin.
    (c) Prothrombin.
    (d) Fibrinogen.

6. Deficiency of vitamin K is rare in adults because:
   (a) Bacteria make vitamin K in the large intestine.
   (b) Vitamin K is common in food.
   (c) Vitamin K is reused in a conservation cycle.
   (d) All of the above.

7. Newborn infants may need:
   (a) Extra green vegetables and oils.
   (b) An injection of vitamin K or oral vitamin K.
   (c) No extra vitamin K.
   (d) An injection of vitamin D.

8. The Dietary Reference Intake (DRI)) of vitamin K for an adult woman is:
   (a) 90 micrograms.
   (b) 90 milligrams.
   (c) 120 micrograms.
   (d) 120 milligrams.

9. Vitamin K is abundant in:
   (a) Red meat.
   (b) Dairy products.
   (c) Leafy green vegetables.
   (d) Fish.

10. Large doses of vitamin K are:
    (a) Very toxic.
    (b) Toxic.
    (c) Slightly toxic.
    (d) Not toxic.

Test: Part Two

Do not refer to the text when taking this test. A good score is at least 18 (out of 25 questions) correct. Answers are in the back of the book. It's best to have a friend check your score the first time, so that you won't memorize the answers if you want to take the test again.

1. The storage form of vitamin A in the liver is:
   (a) Carotenoids.
   (b) Retinol.
   (c) Retinyl palmitate.
   (d) Retinal.

2. Beta-carotene can be split in two to yield two molecules of:
   (a) Retinal.
   (b) Retinyl esters.
   (c) Retinoic acid.
   (d) Retinol.

3. Vitamin A is needed for:
   - (a) Color vision.
   - (b) Vision in bright daylight.
   - (c) Vision in low light.
   - (d) All of the above.

4. Severe deficiency of vitamin A can cause:
   - (a) Blindness in children.
   - (b) Fuzzy vision in adults.
   - (c) Cataracts.
   - (d) Conjunctivitis.

5. When beta-carotene in supplements is converted to vitamin A:
   - (a) It provides half as much as the same amount of retinol.
   - (b) It provides twice as much as the same amount of retinol.
   - (c) It provides the same amount as retinol.
   - (d) It provides one-twelfth as much as the same amount of retinol.

6. Excess preformed vitamin A from animal products or supplements during pregnancy are known to cause:
   - (a) Yellow skin.
   - (b) Rickets.
   - (c) Birth defects.
   - (d) Scurvy.

7. This form of vitamin D is made in the skin:
   - (a) Calcidiol.
   - (b) Cholecalciferol.
   - (c) Calcitriol.
   - (d) Ergocalciferol.

8. This form of vitamin D is a powerful hormone:
   - (a) Calcidiol.
   - (b) Cholecalciferol.
   - (c) Calcitriol.
   - (d) Ergocalciferol.

9. Enough sun can be obtained to produce adequate vitamin D:
   (a) In ten minutes on the face and hands three times weekly.
   (b) In thirty minutes of sun on the face, chest, and arms with SPF 8 sunscreen once weekly.
   (c) Through windows in the winter.
   (d) Supplemental vitamin D must always be taken.

10. Calcidiol is transformed to calcitriol in:
    (a) The liver.
    (b) The kidneys.
    (c) The skin.
    (d) The lungs.

11. Regulation of the active form of vitamin D is handled by:
    (a) The stomach.
    (b) The parathyroid gland.
    (c) The thyroid gland.
    (d) The liver.

12. Which is NOT a risk factor for vitamin D deficiency?
    (a) Being institutionalized.
    (b) Elderly people.
    (c) Breast-fed infants who do not receive sunlight.
    (d) Skin which is light in color.

13. The form of vitamin E that the body prefers to use:
    (a) Alpha-tocotrienol.
    (b) Gamma-tocopherol.
    (c) Alpha-tocopherol.
    (d) Gamma-tocotrienol.

14. Vitamin E neutralizes free radicals in:
    (a) Bones and cartilage.
    (b) Cell membranes and LDL.
    (c) Hair and skin.
    (d) The gallbladder.

15. Vitamin E:
    (a) Is built into LDL for protection from oxidation that can lead to clogged arteries.
    (b) Makes bones stronger.
    (c) Prevents rickets.
    (d) All of the above.

16. To be effective, vitamin E supplements should be taken with meals because:
    (a) It is easier to remember to take supplements at mealtime.
    (b) Fats in food trigger the absorption of vitamin E.
    (c) At bedtime, fatigue prevents absorption.
    (d) Carbohydrates help with absorption of vitamin E.

17. Vitamin E, in amounts above 40 times the RDA, can cause:
    (a) Weak bones.
    (b) Increased respiration.
    (c) Increased bleeding tendencies.
    (d) Hearing loss.

18. The food source highest in vitamin E is:
    (a) Sunflower seeds.
    (b) Fruit.
    (c) Vegetables.
    (d) French fries.

19. Which form of vitamin E is NOT synthetic?
    (a) RRR-alpha-tocopherol.
    (b) SRR-alpha-tocopherol.
    (c) The acetate ester form.
    (d) The succinate ester form.

20. Phylloquinone, vitamin K_1, is synthesized by:
    (a) Bacteria.
    (b) The liver.
    (c) Plants.
    (d) The kidneys.

21. The form of vitamin K synthesized by bacteria is:
    (a) Menaquinone.
    (b) Menadione.
    (c) Menatetrenone.
    (d) Phylloquinone.

22. The name of the protein residue that vitamin K converts is:
    (a) Lysine.
    (b) Arginine.
    (c) Glutamic acid.
    (d) Glycine.

23. Vitamin K is needed for this essential clotting factor:
    (a) Prothrombin.
    (b) Thrombin.
    (c) Fibrin.
    (d) Fibrinogen.

24. Osteocalcin:
    (a) The synthesis is controlled by vitamin D.
    (b) The calcium-binding ability is enabled by vitamin K.
    (c) Is made by osteoblast cells in bones.
    (d) All of the above.

25. The highest source of vitamin K is:
    (a) Cold-water fish such as salmon.
    (b) Kale and collards.
    (c) Dairy products, especially milk.
    (d) Vegetable oils.

PART THREE

The Macro Minerals

Introduction to the Macro Minerals

Macro minerals play a substantial role in our lives. They are the building blocks of life. The macro minerals include calcium, phosphorus, sulfur, potassium, sodium, chloride, and magnesium. These minerals are found in the greatest abundance in the diet and are found in the highest amounts in the body.

Calcium, phosphorus, and magnesium play a large role in the formation and structure of bones. Bones function as a storage pool for these minerals from which the rest of the body can draw as needed. Calcium has important functions in addition to bone structure. Calcium is needed for muscular contractions and blood clotting. On the other hand, magnesium is vital for muscle relaxation. A third macro mineral, phosphorus, is an integral part of cell membranes.

Potassium, sodium, and chloride are needed to form the electrolyte solutions that bathe our tissues. These electrolytes balance the amount of water in the cells and in the blood plasma. These minerals also work as blood buffers to help keep the blood neutral in acidity.

Many Americans could use more calcium and magnesium in their diet. However, Americans generally receive too much sodium and chloride in their diet; this can be partly attributed to the large amount of processed foods consumed. Phosphorus and sulfur levels in the American diet seem to be adequate. These macro minerals need to be eaten in balanced amounts in the diet to maintain good health and to prevent disease.

CHAPTER 7

Water and Electrolytes

We need a constant supply of drinking water for many purposes. We can only survive a few days without water. Water and minerals work together to regulate body fluids.

Water is used to transport nutrients and wastes throughout the body. Water acts as a cushion and lubricant in joints and in the eyes. Water is needed to regulate body temperature. Water is important in maintaining normal blood pressure and blood volume. Water balance is crucial for these and other functions. Water balance is influenced by certain water-soluble minerals.

Water's Functions

A solvent and transporter for many vitamins and minerals.
A cushion and lubricant for joints, eyes, spinal cord, and the amniotic sac.
Used to cool the body.
Needed for blood volume and cellular structure.
Cleans the body of wastes.

About 60 percent of the body weight of adults is comprised of water. Although it may seem that fat is more watery, it turns out that lean tissue contains more water. Lean tissue contains three-quarters water, while fat contains about one-quarter water.

Water Output

To maintain just the right water balance, the intake of water must exactly match the output of water. The body loses or disposes of water in several ways, as displayed in Figure 7-1. The kidneys are the main regulators of water. The kidneys account for about half of the total water output. Our skin loses water in two ways. The loss of water by diffusion from skin accounts for about 15 percent of the total water output, even when we are not sweating. Water loss from sweating varies considerably depending on activity, temperature, and humidity. The lungs continuously lose water as water vapor and account for about 12 percent of water output. The lungs lose extra water vapor during exercise and when the humidity is low. And, finally, about 5 percent of water is lost through the feces.

Under normal conditions, the total amount of water in the body is regulated by the kidneys. Inside the kidneys are tiny tubules that can retain more water when necessary. When blood volume falls or blood pressure falls, or the *extra-*

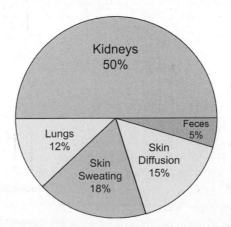

Figure 7-1 How water can leave the body.

cellular fluid becomes too concentrated, the body responds by retaining more water from the urine. The amount of water retained by the kidneys is regulated by hormones, as seen in Figure 7-2. *Antidiuretic hormone* is secreted by the pituitary gland to keep more water in the body. Antidiuretic hormone limits the amount of water lost through the kidneys. Another hormone, *aldosterone*, also directs the kidneys to retain water. Aldosterone is secreted from the adrenal cortex, which is a gland located on the top of each kidney. These two hormones also trigger thirst.

The minimum amount of water that must be excreted each day is about half a quart (500 milliliters). This amount is needed to carry metabolic waste products away in urine. More water than this is needed to allow for sweating, diffusion of water through skin, losses through lungs, and in the feces. Average daily losses run about two and a half quarts (about 2.5 liters). More water is useful to dilute the urine and ensure waste removal. Inadequate water intake has been correlated with increased rates of bladder cancer because of the longer time that the bladder may be exposed to potential carcinogens.

One half-gallon to one gallon is a reasonable range of liquids for most people to drink each day. Hard work in the hot sun greatly increases water needs. Coffee,

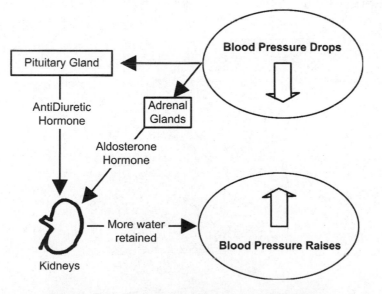

Figure 7-2   Blood pressure and water retention.

tea, sodas, and alcoholic beverages are not good water substitutes as their diuretic action causes losses of about half of the amount taken.

Water Input

The obvious way that we take in water is through drinking water and other beverages, as shown in Figure 7-3. About 55 percent of our water intake is from drinking water and beverages. Food also contains water in varying amounts. Strawberries, watermelon, and broccoli contain about 90 percent water. Bread and cheese contain about 35 percent water. Water from food accounts for about 35 percent of our water intake. Together, water from food and beverages accounts for 90 percent of our water intake.

When food is metabolized for energy in the cells, one byproduct is water. This metabolic water accounts for about 10 percent of our water. When carbohydrates, fats, and protein are burned for energy, their carbon and hydrogen atoms combine with oxygen to produce carbon dioxide and water. Metabolic water may find its way into plasma and can be regarded as intake.

There are mechanisms to adjust the intake of liquids when the total water in the body is high or low. When body fluid stores get low, the mouth feels dry. This thirst encourages drinking to build up body stores of water. When body stores of

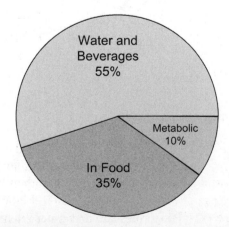

Figure 7-3   How water enters the body.

fluids are high, it is thought that stretch receptors in the stomach signal the brain to stop drinking. Stretch receptors in the bladder signal when it is time to release water. Sensors in the heart also monitor blood volume. Too much water is rarely a problem because it is so easily eliminated. Too little water can lead to dehydration, which can cause weakness and delirium if severe.

Electrolytes

An *electrolyte* solution can be formed when certain mineral salts are dissolved in water. Electrolyte solutions were originally named because of their ability to conduct electricity. Many of the fluids in the body are electrolyte solutions with dissolved minerals. For example, table salt (sodium chloride) can easily dissolve in water to form an electrolyte solution. There are three important areas of the body that contain electrolytes in solution.

Blood plasma is the liquid component of blood. Blood cells are suspended in plasma. Blood plasma is an electrolyte solution that carries blood proteins. A second electrolyte solution is called *intracellular fluid* (*intra* means "within," so "within the cells"). Much of the fluid in the body is inside the cells. The third location of electrolytes in the body is in the *interstitial fluid* (*inter* means "between"). This interstitial fluid bathes and surrounds the cells. The fluid outside of cells is called *extracellular fluid* (*extra* means "outside") and includes both blood plasma and interstitial fluid.

Types of Fluids in the Body

Intracellular Fluid is found inside cells.
Blood Plasma is found inside blood vessels.
Interstitial Fluid is found between the cells.
Extracellular Fluid includes both blood
   plasma and interstitial fluid.

Extracellular fluids, including plasma and interstitial fluid, transport nutrients to cells and transport wastes away from cells. The intracellular fluid acts as a solvent to facilitate the chemical reactions in the cell that maintain life. Intracellular fluid comprises about 40 percent of body weight, as seen in Figure 7-4. Interstitial fluid makes up about 16 percent of body weight. Plasma makes up only about 4 percent of body weight.

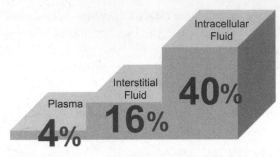

Figure 7-4   Percentage of body weight in plasma, interstitial fluid, and intracellular fluid.

Blood Pressure and Blood Volume

As mentioned, antidiuretic hormone signals the kidneys to retain more water when blood pressure drops. Kidney cells respond to lower blood pressure by releasing the enzyme *renin*, as seen in Figure 7-5. Renin causes the kidneys to retain more sodium. The extra sodium draws more water back from the urine to increase blood volume. Extra dietary sodium can also raise blood pressure by causing more water

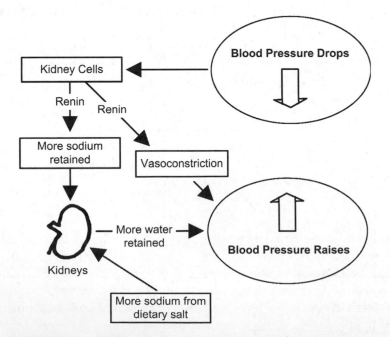

Figure 7-5   Sodium and renin help to control blood pressure.

retention. Renin also triggers *vasoconstriction*, the narrowing of blood vessels. This also raises blood pressure. Renin activates a blood protein, *angiotensin*, which causes the vasoconstriction.

Fluid and Electrolyte Balance in the Cells

Cells regulate their internal mineral balance, which keeps just the right amount of water inside the cells. If there is too much water inside the cells, the cells can burst. With too little water inside the cells, they can collapse. Cells maintain a fluid balance where about two thirds of the water in the body is inside the cells and one third of the water is outside the cells.

Mineral salts, such as sodium chloride (NaCl) dissolve in water. Sodium chloride breaks apart into positively charged sodium ions and negatively charged chloride ions. Water itself has slightly different charges on the different atoms making up the water molecule (H_2O), as shown in Figure 7-6. The two hydrogen

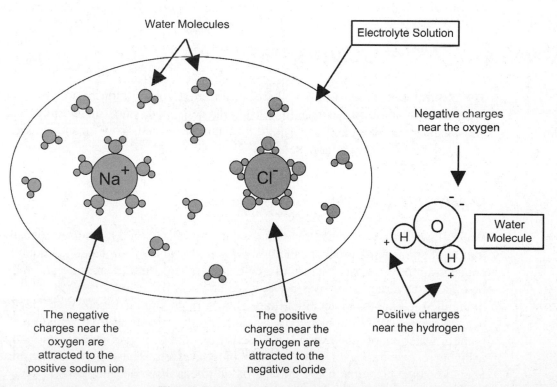

Figure 7-6   How salts dissolve in water.

atoms in water have slight positive charges that are attracted to the negatively charged chloride ions. The water molecules cluster around the chloride ions with the hydrogen sides of the water molecules facing the chloride ion. On the other side of the water molecule, the oxygen atom has a slight negative charge that is attracted to the positively charged sodium ions. The water molecules cluster around the sodium ions with the oxygen sides of the water molecules facing the sodium ion. This is how water dissolves salts.

Ions are charged particles. Positive ions, such as sodium ions are called *cations*. Negatively charged ions, such as chloride, are called *anions*. This is easier to remember if you think of the "t" in cations as being similar to a "+" sign. In electrolyte solutions, the number of positive ions always equals the number of negative ions. Electrolyte solutions in the body are always neutral.

Each cell maintains this electrical neutrality by making sure that if a positive sodium ion leaves the cell, a positive ion, often potassium (K^+), enters the cell at the same time. These ion concentrations are measured in *milliequivalents per liter* (mEq/L). Inside cells, normal values are 202 mEq/L, while the extracellular fluid has a normal value of 155 mEq/L. Some ions have a double charge: for instance, calcium (Ca^{++}) and phosphate ($HPO_4^=$).

Movement of Electrolytes

The electrolyte solution inside cells has a different composition of solutes than the extracellular electrolyte solution. Inside the cells, potassium, magnesium, and phosphate are in the highest concentrations. Outside the cells, sodium and chloride are in the highest concentrations.

OSMOTIC PRESSURE

Water flows across cell membranes and capillary walls toward the higher concentration of solutes in a process called *osmosis*. *Osmotic pressure* tends to equalize concentrations. Osmosis creates a force that is called osmotic pressure and may be offset by hydraulic pressure in a delicate balance. Blood vessels have a higher hydraulic pressure than interstitial fluid. This extra pressure in blood vessels tends to move water out of the blood vessels and into the interstitial fluid. This movement of water out of blood vessels is offset by osmotic pressure. There is normally a higher concentration of solutes inside blood vessels. Thus, osmotic pressure tends to move interstitial water into the blood to equalize concentrations.

TRANSPORT OF IONS ACROSS CELL MEMBRANES

Transport proteins embedded in the cell membrane regulate the flow of positive ions across the cell membrane. Negative ions follow the positive ions to maintain a neutral electrolyte balance. Remember that electrolyte solutions must be electrically neutral. Also, water flows across cell membranes toward the more concentrated solution because of osmosis.

One of the best understood transport proteins is the sodium-potassium pump. Proteins embedded in the cell wall actively pump sodium out of the cell while bringing potassium into the cell, as shown in Figure 7-7. First, potassium binds to the protein on the outside of the cell membrane. The protein then flips over, delivering the potassium into the intracellular fluid. Once inside the cell, the now-available

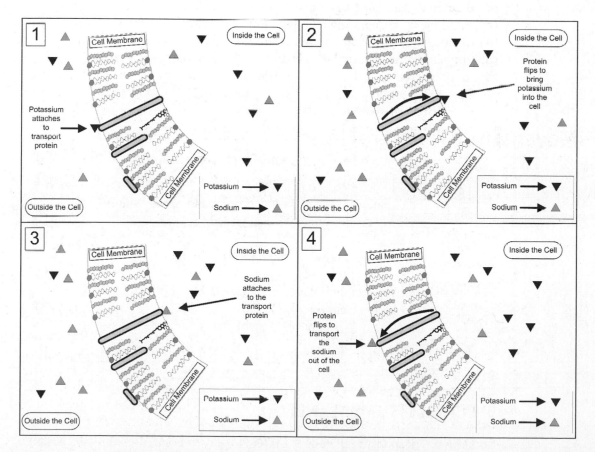

Figure 7-7   Cell membrane transport of sodium and potassium.

protein binds to sodium. Next, the protein flips over again to transport the sodium out of the cell. This kind of active transport requires energy in the form of adenosine triphosphate (ATP).

Fluid and Electrolyte Balance in the Body

The body must have the right amount and proportion of minerals at all times. The gastrointestinal tract and the kidneys work together to regulate minerals in the body. Minerals are recycled back to the stomach and intestines in the form of digestive juices and bile. These minerals, together with the minerals in food, are absorbed or reabsorbed as needed. About two gallons of digestive juices and minerals are recycled each day. Selective reabsorption through the intestines allows regulation of fluid and electrolyte balance.

The kidneys do more than simply adjust the amount of water retained by the body. Kidneys also regulate certain electrolytes. For example, when body stores of sodium get low, the adrenal glands secrete the hormone aldosterone. As you may recall, aldosterone helps the kidneys retain more sodium. When more sodium is retained, the electrolyte balance is maintained when the kidneys release another positive ion, often potassium.

Summary for Fluids and Electrolyte Balance

Water intake must equal water output.
The kidneys regulate blood pressure and blood volume.
Electrolytes are mineral ions that help regulate fluid balance.
Electrolytes are found in the plasma, in interstitial fluids, and inside the cells.
Electrolyte solutions are neutral with equal positive and negative charges.
Blood acidity is controlled by blood buffers, the kidneys, and the lungs.

ELECTROLYTE IMBALANCE

Sometimes, despite the excellent systems used to control electrolyte balance, the body's mineral and fluid balances do get disrupted. It is not uncommon for heavy sweating to deplete fluids and electrolytes. Wounds can also throw off electrolyte balance through loss of blood. Diarrhea or vomiting can purge large amounts of fluids and minerals in a short time. Sodium and chloride losses are the most common because they are the minerals in the highest concentration in extracellular fluids.

Oral rehydration therapy is often administered when severe diarrhea threatens the health of malnourished children. Oral rehydration therapy is one cup of pure water with a teaspoon of sugar and a little salt. A zinc supplement added to oral rehydration therapy aids the therapy. This therapy can be helpful until the child is rehydrated and strong enough to recover with normal water and food.

The Acid-Alkaline Balance

The blood must be maintained in a narrow range of acidity. Acidity is measured on a *pH* (potential of hydrogen) scale. Normal values for blood are between 7.35 and 7.45 on the pH scale. Water is used as a reference as pH 7, which is considered neutral. Values higher than pH 7 are considered *alkaline* (also known as *base*). Baking soda has an alkaline pH of 9. Values lower than pH 7 are considered acid. Vinegar has an acidic pH of 3. If the blood deviates from its narrow range of normal acidity, delicate proteins can be damaged. Enzymes need normal acidity to perform their functions and are vital to life. In hospitals the acidity of intravenous solutions is carefully adjusted for proper pH.

The concentration of hydrogen ions (H^+) determines the acidity of a solution. The higher the concentration of hydrogen ions, the higher the acidity. Hydrogen ions are generated during normal energy metabolism in the cells. Other acids are also generated during normal energy metabolism. These acids must be neutralized. There are three different systems in the body that work together to normalize acidity. The kidneys, the lungs, and pH buffers in the blood keep our blood at the correct, neutral acidity.

**Three Systems to Maintain
Blood Acid-Alkaline Balance**

Kidneys release acidity.
Lungs release carbon dioxide.
Blood buffers neutralize blood.

The kidneys are the most important regulators of acid-alkaline balance. The kidneys decide which ions to keep and which ions to reject. The acidity of urine is adjusted so that the acidity of blood stays at the proper level.

Protein in excess of the adult daily requirement of 46 to 56 grams is burned for energy. When this excess protein is burned for energy, metabolic acids are formed. This puts an extra strain on the blood buffering systems. Vegetables, which leave an alkaline residue, help to offset excess metabolic acids.

Lungs also regulate the acidity of blood. During normal metabolism, carbon dioxide is formed. This carbon dioxide forms carbonic acid in the blood. This acidic influence on the blood must be neutralized. Carbon dioxide is released during respiration. If acid builds up in blood, breathing accelerates to release more carbon dioxide, thus lowering the carbonic acid levels in the blood. Conversely, if blood acidity is a little too low, respiration slows to allow more carbonic acid to build up again to regain neutral acidity in the blood. The lungs release an average of thirty liters (about eight gallons) of carbonic acid each day to keep the blood neutral.

> Lungs release enough carbon dioxide each day to neutralize eight gallons of carbonic acid.

BLOOD BUFFERS

Blood acidity is also regulated by blood buffers. These buffers come in pairs that work together. One pair of buffers consists of bicarbonate (alkaline) and carbonic acid (acid). Buffers can be used up if a continuous supply of extra acid or alkaline substances is added to the blood.

It is more common for excess acidity to be a problem since many metabolic wastes are acidic. If too much acid enters the blood, the bicarbonate may become depleted. Respiration may be temporarily increased to release more carbonic acid and restore the ratio of bicarbonate to carbonic acid.

Another pair of blood buffers consists of *phosphoric acid* and *dihydrogen phosphate*. This pair of blood buffers also consists of an acid-alkaline pair. Proteins in blood also can act as blood buffers by donating or accepting positive hydrogen ions.

A healthy body does an excellent job of adjusting fluids, acidity, and the balance of minerals.

Quiz

Refer to the text in this chapter if necessary. A good score is at least 8 correct answers out of these 10 questions. The answers are listed in the back of this book.

1. Which statement is NOT true?
   (a) 60 percent of the body weight of adults is comprised of water.
   (b) Fat contains about 90 percent water.
   (c) Lean tissue contains three-quarters water.
   (d) Fat contains about one-quarter water.

2. The body loses the least amount of water through:
   (a) The lungs.
   (b) Skin diffusion.
   (c) The feces.
   (d) The kidneys.

3. Which statement is NOT true?
   (a) The minimum amount of water needed daily by kidneys is one half-quart.
   (b) The minimum amount of water to meet needs is two and one half quarts.
   (c) The amount of water most people need each day is one half-gallon to one gallon.
   (d) The amount of water needed each day can be met by one half-gallon of tea.

4. An electrolyte solution can be formed when:
   (a) A mineral salt is dissolved in water.
   (b) Fats are added to water.
   (c) Calcium is added to bones.
   (d) Mineral salts are removed from water.

5. Which one is NOT known to be an electrolyte solution:
   (a) Blood plasma.
   (b) Interstitial fluid.
   (c) Extragalactic fluid.
   (d) Intracellular fluid.

6. When table salt is dissolved in water, the sodium is attracted to:

    (a) The hydrogen side of the water molecule.

    (b) The oxygen side of the water molecule.

    (c) The chloride ions.

    (d) None of the above.

7. The sodium-potassium pump:

    (a) Pumps potassium into the cell.

    (b) Pumps sodium into the cell.

    (c) Pumps potassium into the plasma.

    (d) Pumps manganese into the plasma.

8. What quantity of digestive juices and bile are recycled each day?

    (a) One quart.

    (b) One half-gallon.

    (c) One gallon.

    (d) Two gallons.

9. Electrolyte solutions inside the cell are:

    (a) Negatively charged.

    (b) Positively charged.

    (c) Electrically neutral.

    (d) More positively charged than negatively charged.

10. To maintain acid-alkaline balance:

    (a) Kidneys release acidity.

    (b) Lungs release carbon dioxide.

    (c) Blood buffers neutralize blood.

    (d) All of the above.

CHAPTER 8

The Electrolyte Minerals
Sodium, Chloride, and Potassium

Minerals—an Introduction

Minerals in nutrition fall into two categories. Major minerals (sometimes called *macro minerals*) are needed in greater amounts in the diet and are found in greater amounts in the body, as seen in Figure 8-1. *Trace minerals* are needed in smaller amounts in the diet and are found in lesser amounts in the body, as seen in Figure 8-2. All of the major minerals in Figure 8-1 and the trace minerals in Figure 8-2 are essential for nutrition.

The key to proper mineral nutrition is balance. Minerals should not be eaten in amounts that greatly exceed needs. Some minerals are toxic in excessive amounts. Some minerals, when taken in excess, induce a relative deficiency of other minerals. For example, excessive sodium causes calcium losses. The body needs every one of the nutritional minerals. Deficiency of even one mineral should be avoided.

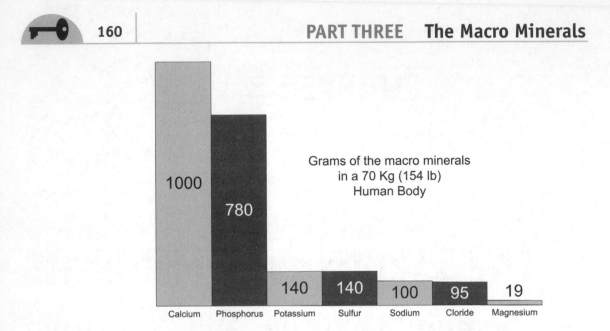

Grams of the macro minerals
in a 70 Kg (154 lb)
Human Body

Figure 8-1   Macro minerals in the human body in grams.

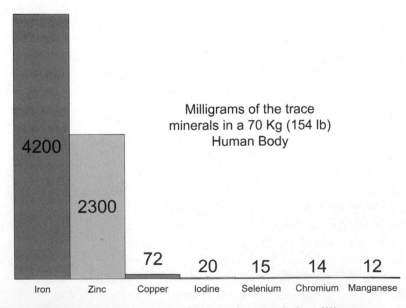

Milligrams of the trace
minerals in a 70 Kg (154 lb)
Human Body

Figure 8-2   Some important trace minerals in milligrams.

All minerals come from the soil, whether directly from plants, or indirectly from animals that eat plants. The minerals must exist in the soil in order for them to be absorbed by plants. Also, microbes and rootlet fungi in the soil must be present and healthy to enable the plants to absorb minerals. Some agricultural regions are low in certain minerals. Certain agricultural practices can kill some of the soil organisms that assist the uptake of minerals. Organically grown plants may have a more complete spectrum of minerals.

Minerals are inorganic elements. Minerals are not destroyed by heat in cooking or processing. Minerals are, however, susceptible to being leached out into cooking water that is discarded. Minerals may combine with other compounds in the body, but they retain their unique identity and do not change.

> Phytates and Oxalates can Limit Mineral Absorption.
> Phytates occur in grains and beans.
> Oxalates occur in some green vegetables.

Some of the minerals in food are unavailable for absorption. These unavailable minerals may be bound with *oxalates* or *phytates* that prevent their absorption. Phytates are found in grains and beans and limit the absorption of some of the minerals. Oxalates are found in such vegetables as spinach and beet greens. These foods still provide excellent mineral nutrition, despite a slight reduction in mineral bioavailability.

Sodium

Salt is made of sodium and chloride—salt is essential for life. The chemical symbol for sodium, *Na*, is derived from the Latin name for sodium, *natrium*. Most people enjoy the taste-enhancing flavor of salt. Less than 10 percent of salt intake is from the salt in unprocessed foods such as fruits and vegetables. Three-quarters of the salt most people eat is hidden in processed foods. Many processed foods have a high sodium content without tasting salty. About 10 percent of normal salt intake is from salt added in the kitchen or at the table.

> Three-quarters of the salt most people
> eat is hidden in processed foods.

Most natural foods start out with an abundance of potassium and very little sodium. During food processing, this balance gets reversed. Processed foods contain less potassium and excess sodium. An apple starts with only one milligram of sodium. The same weight of apple pie has 266 mg of sodium, as seen in Figure 8-3.

One teaspoon of salt weighs about five grams, and contains 2000 mg sodium (40 percent sodium). The maximum safe amount of sodium per day is set at 2400 mg sodium, just over a teaspoon of salt. Few people stay below the safe level.

Sodium (a cation, Na^+) and chloride (an anion, Cl^-) are the most abundant ions in the fluids outside of cells (extracellular fluid). Sodium is important for maintaining blood pressure and fluid balance. Sodium retention in the kidneys can result in increased water retention, which can result in increased blood pressure. The kidneys remove all sodium from the blood and then add back just the right amount of sodium to the blood. To adjust acid-alkaline balance and lower blood acidity, the kidneys can excrete hydrogen ions (H^+) and exchange them for sodium ions (Na^+).

Sodium is absorbed directly from the intestinal tract. Sodium travels freely in blood and interstitial fluid. Excess sodium causes thirst, which triggers extra water intake. This extra water flushes the extra sodium out through the kidneys.

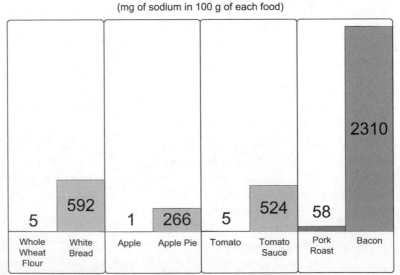

Figure 8-3   Sodium is increased in food processing.

CELL MEMBRANE POTENTIAL

The sodium-potassium pumps in cell membranes maintain concentrations of sodium and potassium, which are very different inside the cell than they are outside the cell. Sodium concentrations are 10 times higher outside cells than inside cells. Potassium concentrations are thirty times higher inside cells than outside cells, as shown in Figure 8-4. The active pumping of sodium and potassium requires energy. It has been estimated that about 30 percent of the energy used in the body at rest is used to maintain this pumping action. The different concentrations of these minerals create an electrochemical gradient known as the *membrane potential*. The control of cell membrane potential is critical for heart function, nerve impulse transmission, and muscle contraction.

DEFICIENCY OF SODIUM

Sodium deficiency does not normally result from inadequate dietary intakes. Sodium deficiency is called *hyponatremia* (*hypo-* means "low," *natrium* means "sodium," and *-emia* means "in blood"). In rare cases, excessive water intake can cause low sodium levels in the blood. Hyponatremia can be caused by prolonged, excessive sweating, prolonged vomiting or diarrhea, or the use of some diuretics. Symptoms of hyponatremia include headache, muscle cramps, fainting, fatigue, and disorientation. Hyponatremia is something to watch for with intense sport competitions that last for many hours.

Certain drugs can cause lowered sodium levels. These drugs include some diuretics, ibuprofen, naproxen, Prozac, and Elavil.

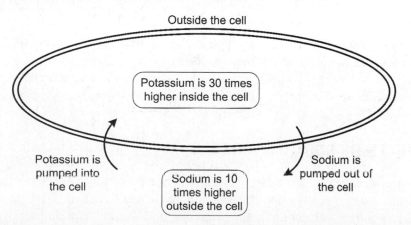

Figure 8-4   Sodium and potassium concentrations inside and outside of the cell.

RECOMMENDED INTAKE OF SODIUM

Adequate Intake levels (AI) have been set by the Food and Nutrition Board of the Institute of Medicine for sodium. These levels are based on the amount of salt needed to replace losses with moderate sweating in average people. The AI for children and adults ranges from 1.0 to 1.5 grams per day. The tolerable upper intake levels for sodium range from 2.4 to 3.8 per day for children and adults. In spite of these upper levels, the dietary salt intake in the United States averages 10 grams per day for adult men and seven grams per day for adult women—more than twice the tolerable upper intake levels.

Summary for Sodium

Main functions: maintains blood pressure and fluid balance, assists muscle contraction, and assists nerve impulse transmission.

Adequate Intake: for adults and children, it ranges from 1 to 1.5 grams per day.

Toxicity is rare. Excess intake can increase risk of high blood pressure.

Tolerable upper intake level is set at 3.8 g for adults. Over age 70 it is 3.0 g daily.

Deficiency is from excessive losses such as losses from excessive sweating.

Healthy sources: unprocessed fruit, vegetables, whole grains, and unsalted nuts.

Unhealthy sources: processed food often contains too much sodium.

Forms in the body: free sodium ion and bound to chloride.

Evidence is consistent that diets high in potassium (over 4.5 grams per day) and low in sodium (under 6 grams per day) decrease the risk of high blood pressure and the risk of strokes and heart attacks. This is easiest to achieve with a diet made up predominantly of unprocessed fresh fruit, vegetables, and whole grains.

SOURCES OF SODIUM

Three-quarters of the salt consumed by Americans is hidden in processed food. Only one cup of macaroni and cheese has a whole day's dose of salt. Unprocessed food such as fruit, vegetables, grains, legumes, and unsalted nuts and seeds are low in sodium; please refer to Graph 8-1.

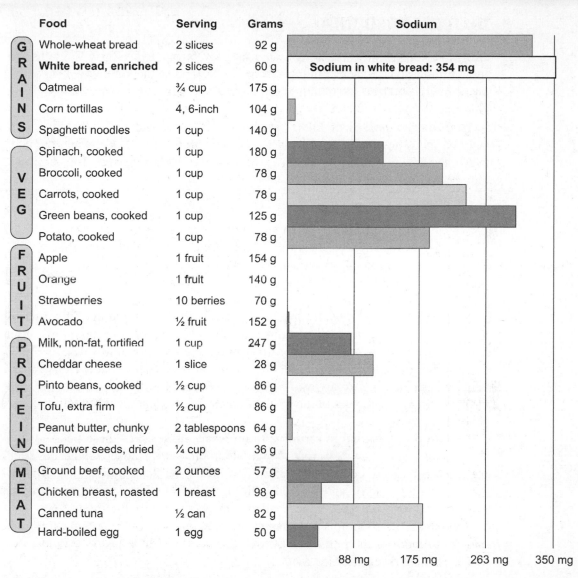

Graph 8-1 Sodium content of some common foods.

TOXICITY OF SODIUM

Excessive intakes of salt lead to increasing levels of extracellular fluids as the sodium draws the fluids from inside the cells. As long as the kidneys are functioning normally and there is enough water, the extra sodium is flushed out through the kidneys.

High dietary intakes of highly salted food such as pickles, salted fish, and smoked foods have been tentatively linked to increases in stomach cancer. There is an increase in calcium loss with higher salt intakes, making excess salt a risk factor in osteoporosis.

> Reducing salt in the diet has been shown to slightly lower the risk of stroke, high blood pressure, and heart attack.

Reducing salt intake by six grams per day results in an average lowering of the systolic blood pressure by about four mm of mercury in people with high blood pressure. This may not seem like much, but over the entire U.S. population, this slight reduction could reduce the prevalence of heart attacks by 6 percent and the risk of strokes by 15 percent. With one million heart attacks and 600,000 strokes each year in the United States, a 6 to 15 percent reduction could save many lives.

Certain people are more sensitive to high salt intakes. For these people, high salt intakes result in an extra risk of high blood pressure. Salt sensitivity has been reported to be more common in obese and insulin-resistant individuals, African Americans, the elderly, and women with high blood pressure. For these people, lowering salt intake to the recommended levels is especially important.

To summarize, sodium is a needed nutrient, but most people take in levels high enough to increase the risk of cardiovascular disease. Much of the sodium consumed is hidden in processed food.

Chloride

Chlorine (chemical symbol *Cl*) is a poisonous, greenish-yellow gas in the halogen family of elements. Chlorine is used to disinfect water and is used as bleach. Chlorine combines easily with sodium, hydrogen, and many other elements.

When combined with sodium or hydrogen, chlorine forms a stable ion called *chloride* (Cl^-). The chloride ion has a single negative charge. Table salt is sodium chloride.

Chloride is an essential nutrient. Chloride is the principal anion (negatively charged ion) in the extracellular fluid. Chloride is an electrolyte that works with sodium to maintain the cellular membrane potential. These electrolytes contribute to the maintenance of charge differences across cellular membranes.

> Hydrochloric acid in the stomach is made with chloride.

Chloride is a component of hydrochloric acid (HCl). Hydrochloric acid is an important component of gastric juice, which aids in the digestion of certain nutrients. Hydrochloric acid contributes to the acidity of gastric juice. If gastric juice is lost through vomiting, enough acid can be lost to affect the acid-alkaline balance of the body.

CHLORIDE DEFICIENCY

Deficiency of chloride is rare because it is found in most diets. Temporary deficiency of chloride can be induced by heavy sweating, vomiting, or diarrhea. Deficiency can be relieved by normal food and water. In extreme cases, rehydration therapy may be needed.

Summary for Chloride

Main functions: maintains blood pressure and fluid balance.
Minimum requirement: 750 mg per day.
Toxicity is rare. Excess intake can increase risk of high
   blood pressure.
Deficiency is from excessive losses such as losses from
   vomiting.
Unhealthy sources: processed food often contains too
   much sodium and chloride.
Forms in the body: bound to sodium as salt or to hydrogen
   as hydrochloric acid.

CHLORIDE SOURCES

Chloride is abundant in food, commonly in the form of salt (sodium chloride). An RDA has not been set for chloride. However, a minimum requirement of 750 mg per day has been set. As mentioned with sodium, chloride is overly abundant in processed food.

TOXICITY OF CHLORIDE

Chloride is not toxic. The only noted occurrences of excess chloride were due to water deficiencies. The normal dietary salt excesses of the average American diet contribute to increased risk of high blood pressure. The excessive intake of sodium chloride can lead to an increase in blood volume.

Potassium

Potassium is an electrolyte mineral that is essential for nutrition. The chemical symbol for potassium is the letter K, named after the Latin word *kalium*. The word kalium comes from the Arabic word for alkali, which means cooked ashes. Potassium dissolves into charged particles (ions) in watery fluids. Cells need a high concentration of potassium inside cells to function normally. As mentioned, potassium concentrations are about 30 times higher inside cells than in extra-cellular fluid.

Cells work hard using a *sodium-potassium ATPase* pump to keep the potassium concentration high inside the cells. The *ATP* in the name of this pump indicates that it requires energy. This pump maintains an electrochemical gradient known as the *membrane potential* across the cellular membranes.

POTASSIUM AND ENERGY PRODUCTION

Potassium is needed by an enzyme called *pyruvate kinase*. This enzyme is used to break down carbohydrates for energy production in the cell. Pyruvate kinase is also involved in synthesizing glucose in the liver. One of the signs of potassium deficiency is fatigue, which may be caused by a lack of potassium in the pyruvate kinase enzyme, leading to a lack of energy.

OSTEOPOROSIS

Diets rich in fruits and vegetables have been found to lower the risk of osteoporosis. There are many nutrients in fruits and vegetables, including potassium and calcium. Potassium in food and in supplements decreases calcium loss through the kidneys, which increases bone formation and lowers bone loss.

Potassium-rich foods, such as fruits and vegetables, increase the available acid buffers in blood, especially bicarbonate. Bicarbonate reduces blood acidity. American diets tend to be low in foods that leave an alkaline residue in the body, such as unprocessed fruit and vegetables. At the same time, American diets tend to be high in foods that leave extra acid residues in the body, such as meat, fish, eggs, and cheese. Normal metabolism also leaves acid residues in the blood that need to be buffered.

If the amount of potassium-rich fruits and vegetables eaten is not sufficient to produce enough alkalinity to buffer blood acids, the body has the ability to remove calcium from bones. This calcium helps neutralize the blood, but leaves the bones depleted in calcium. This increases the risk of osteoporosis.

Increasing the amount of potassium-rich fruits and vegetables in the diet helps to preserve calcium in bones. The calcium can then stay in the bones because the extra potassium in fruits and vegetables buffers blood acidity. In support of this theory, potassium bicarbonate supplementation has been found to decrease urinary acid excretion and to decrease urinary calcium excretion. Of course, it is best to obtain potassium from the diet. Decreased urinary calcium also helps lower the risk of kidney stones.

POTASSIUM DEFICIENCY

Potassium deficiency rarely results from a low dietary intake. Potassium deficiency is called *hypokalemia* (*hypo*- means low and *kalium* is Latin for potassium). The causes of hypokalemia are related to excessive losses of potassium from the body. This can be caused by prolonged vomiting, certain drugs, and some forms of kidney disease. Large amounts of licorice can lower potassium levels. Licorice contains glycyrrhizic acid, which has effects similar to those of aldosterone. Aldosterone is a hormone that increases urinary excretion of potassium.

The symptoms of hypokalemia can result from alterations in membrane potential and lack of potassium for energy production. Symptoms include fatigue, muscle weakness, bloating, and intestinal sluggishness. Severe hypokalemia can result in muscular paralysis or abnormal heart rhythms that can be fatal.

FOOD SOURCES OF POTASSIUM

Fruits and vegetables are the richest sources of potassium; please refer to Graph 8-2. People who eat large amounts of fruits and vegetables have a high potassium intake of about 10 grams daily (10,000 mg). Average dietary intake in the United States is two to three grams daily. It is safe and healthy to eat high amounts of potassium in the diet.

As food becomes more processed, the intake of potassium decreases while the intake of salt increases. In primitive cultures, salt intake is seven times lower than potassium levels. In America today, salt intake is three times higher than potassium intake, as shown in Figure 8-5.

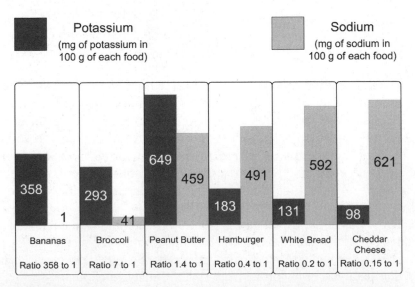

Figure 8-5   Potassium and sodium ratios in some common foods.

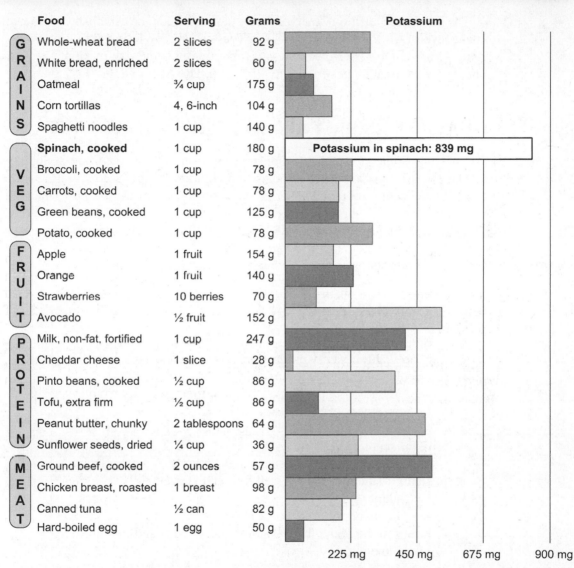

Food	Serving	Grams	Potassium

GRAINS
Whole-wheat bread	2 slices	92 g	
White bread, enriched	2 slices	60 g	
Oatmeal	¾ cup	175 g	
Corn tortillas	4, 6-inch	104 g	
Spaghetti noodles	1 cup	140 g	

VEG
Spinach, cooked	1 cup	180 g	Potassium in spinach: 839 mg
Broccoli, cooked	1 cup	78 g	
Carrots, cooked	1 cup	78 g	
Green beans, cooked	1 cup	125 g	
Potato, cooked	1 cup	78 g	

FRUIT
Apple	1 fruit	154 g	
Orange	1 fruit	140 g	
Strawberries	10 berries	70 g	
Avocado	½ fruit	152 g	

PROTEIN
Milk, non-fat, fortified	1 cup	247 g	
Cheddar cheese	1 slice	28 g	
Pinto beans, cooked	½ cup	86 g	
Tofu, extra firm	½ cup	86 g	
Peanut butter, chunky	2 tablespoons	64 g	
Sunflower seeds, dried	¼ cup	36 g	

MEAT
Ground beef, cooked	2 ounces	57 g	
Chicken breast, roasted	1 breast	98 g	
Canned tuna	½ can	82 g	
Hard-boiled egg	1 egg	50 g	

225 mg   450 mg   675 mg   900 mg

Graph 8-2 Potassium in some common foods.

The Food and Nutrition Board of the Institute of Medicine established adequate intake levels (AI) for potassium in 2004. These levels are designed to supply adequate potassium to lower blood pressure and minimize the risk of kidney stones. The AI for children ranges from 3 to 4.5 grams. The AI for adults is 4.7 grams.

POTASSIUM TOXICITY AND POTASSIUM SUPPLEMENTS

Potassium is limited to 99 mg in supplemental form in the United States. Supplemental potassium is available as potassium chloride, bicarbonate, citrate, aspartate, gluconate, and orotate. Higher amounts of potassium can be prescribed for potassium depletion; this requires careful monitoring of blood potassium levels. There is a potential for serious side effects with higher levels of potassium supplementation. Symptoms of *hyperkalemia* (excess potassium) include weakness and tingling sensations and can lead to cardiac arrest in extreme cases.

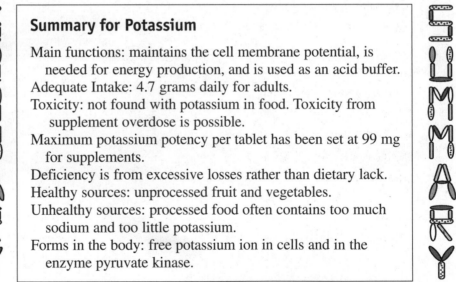

Summary for Potassium

Main functions: maintains the cell membrane potential, is needed for energy production, and is used as an acid buffer.

Adequate Intake: 4.7 grams daily for adults.

Toxicity: not found with potassium in food. Toxicity from supplement overdose is possible.

Maximum potassium potency per tablet has been set at 99 mg for supplements.

Deficiency is from excessive losses rather than dietary lack.

Healthy sources: unprocessed fruit and vegetables.

Unhealthy sources: processed food often contains too much sodium and too little potassium.

Forms in the body: free potassium ion in cells and in the enzyme pyruvate kinase.

The most common side effects of high doses of potassium supplements are gastrointestinal disturbances such as nausea, vomiting, and diarrhea. It is best to take high doses of potassium in a microencapsulated form and with meals to reduce side effects.

To summarize, potassium is very useful in balancing acidity in the body. Many Americans could benefit by eating more unprocessed foods, which are rich in potassium.

Quiz

Refer to the text in this chapter if necessary. A good score is at least 8 correct
answers out of these 10 questions. The answers are listed in the back of this book.

1. Which mineral is a trace mineral?
    (a) Calcium.
    (b) Phosphorus.
    (c) Iron.
    (d) Magnesium.

2. Phytates and oxalates limit mineral absorption and are found in:
    (a) Legumes.
    (b) Grains.
    (c) Some green vegetables.
    (d) All of the above.

3. What percentage of their salt intake do Americans get from processed
   food?
    (a) 10 percent.
    (b) 25 percent.
    (c) 50 percent.
    (d) 75 percent.

4. Sodium deficiency is seen with:
    (a) Excessive sweating.
    (b) Excessive salt intake.
    (c) Lack of sodium in the diet.
    (d) Excessive potassium in the diet.

5. The tolerable upper intake level for sodium intake for men is 3.8 grams.
   How much sodium does an average American man eat?
    (a) Three grams.
    (b) Six grams.
    (c) Ten grams.
    (d) Twelve grams.

6. Dietary lack of potassium can cause:

  (a) Sodium deficiency.

  (b) Chloride deficiency.

  (c) Calcium loss from bones.

  (d) Stronger bones.

7. Potassium deficiency is called:

  (a) Hypokalemia.

  (b) Hyperkalemia.

  (c) Hyponatremia.

  (d) Hypernatremia.

8. Which type of food has abundant potassium?

  (a) Meats.

  (b) Vegetables.

  (c) Eggs.

  (d) Fish.

9. Ten grams of potassium from food each day is an excellent amount. What is the average amount of potassium in American diets?

  (a) Eight to ten grams.

  (b) Five to eight grams.

  (c) Three to five grams.

  (d) Two to three grams.

10. Potassium supplements in the United States are limited in potency to:

  (a) Under 100 mg.

  (b) Under 500 mg.

  (c) Under one gram.

  (d) Under four grams.

CHAPTER 9

Calcium
The Bone Builder

Calcium is the most abundant mineral in the human body. The chemical symbol for calcium is Ca. In the elemental makeup of the human body, calcium ranks fifth after oxygen, carbon, hydrogen, and nitrogen. Calcium makes up about two percent of adult body weight. Calcium in the body rises from an average of 24 grams at birth to about 1300 grams (about three pounds) at maturity. During the 20 or so years of growth, an average of 180 mg of calcium needs to be added to the bones each day. In the maximum growth period of life (ages 10 to 17), 300 mg of calcium need to be added to the bones each day.

Calcium is an essential nutrient that has a vital role in nerve and muscle function. Calcium assists with enzyme processes and blood clotting. One of the most obvious roles of calcium in the body is to provide rigidity to bones. The mineral component of bones consists mainly of *hydroxyapatite* crystals, which are made of calcium and phosphate. The hydroxyapatite crystals are embedded in collagen.

Most of the calcium in the body, about 99 percent, is found in the bones and teeth. Only one percent is found in the blood and soft tissue. Calcium must be maintained in the bloodstream within a narrow range of concentrations. This is so important to survival that the body will remove calcium from bones to keep the blood concentration constant.

Bone Remodeling

Bone tissue is constantly being dissolved and replaced in the body in a process called *bone remodeling*. Bone cells called *osteoclasts* begin the process of bone remodeling by dissolving bone, as diagrammed in Figure 9-1. This is called bone *resorption*. New bone is then synthesized by bone cells called *osteoblasts* to replace the resorbed bone. When bone formation exceeds resorption, bone is growing. When bone formation is slower than bone resorption, osteoporosis may result. Ideally, bone formation matches bone resorption in adults.

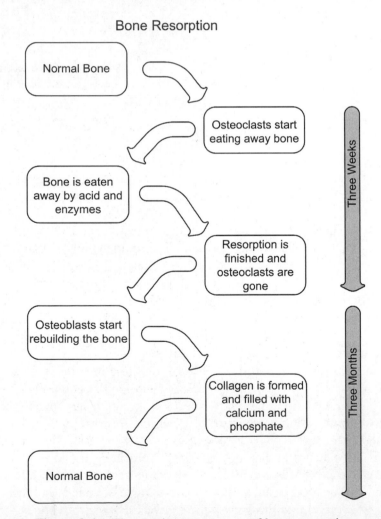

Figure 9-1   The continuous process of bone resorption.

Calcium and Muscle Contraction

Calcium plays a vital role in muscle contraction. Skeletal muscle cells have *calcium channels* in their cell membranes that respond to nerve impulses for contraction. When a nerve gives a signal to a muscle cell to contract, the calcium channels in the cell membrane open, allowing a few calcium ions into the muscle cell. Once inside the cell, the calcium ions attach to activator proteins in the cell, as seen in Figure 9-2. The activator proteins in the cell trigger the release of large numbers of calcium ions from storage inside the cell. These released calcium ions bind to a protein, *troponin-c*, leading to muscle contraction. To assist the contraction, calcium also binds to another protein, *calmodulin*. Calmodulin helps release blood

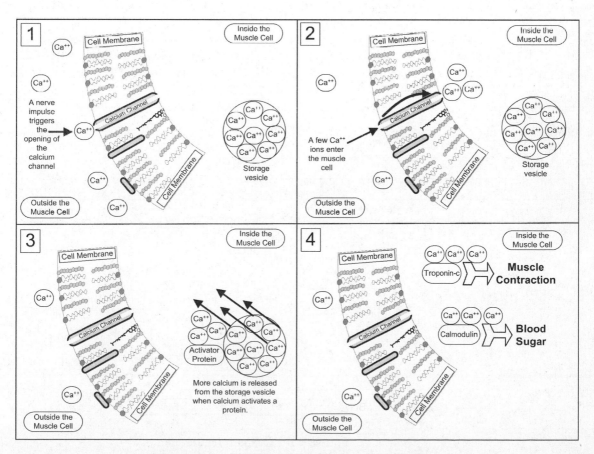

Figure 9-2   Calcium enables muscle contractions.

sugar from storage to fuel the contraction. Calcium channels are used by many cells for cell signaling.

How Calcium Is Regulated in the Blood

Calcium concentration in the blood must be maintained within tight limits. The parathyroid gland and the active form of vitamin D, calcitriol, work together to maintain these calcium levels. When the parathyroid gland senses that blood calcium is low, it releases parathyroid hormone, as diagrammed in Figure 9-3. The parathyroid hormone triggers the kidneys to release calcitriol and reabsorb more calcium to boost blood calcium levels. Kidney resorption is the quickest way to raise blood calcium levels. If more blood calcium is needed, the kidneys release more calcitriol, which causes increased absorption of calcium from the intestines. If still more calcium is needed, calcitriol forces bones to give up some of their

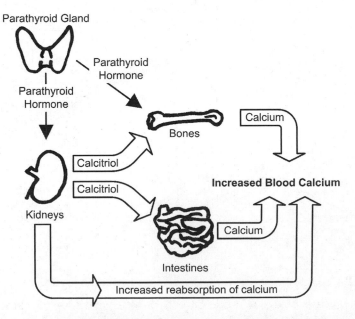

Figure 9-3   Calcium regulation.

calcium into the bloodstream. Calcitriol removes calcium from bones by activating the osteoclast cells in the bones.

Calcium's Roles in the Body

Calcium is needed by bones and teeth for strength.

Calcium enables muscle contraction.

Calcium helps the blood vessels relax and constrict.

Calcium is needed for nerve impulse transmission.

Calcium is used in the secretion of hormones such as insulin.

Calcium is needed by a number of proteins and enzymes.

The binding of calcium ions is required for blood clotting.

Adequate calcium may be helpful in reducing high blood pressure.

Calcium may help prevent intestinal cancer.

Deficiency of Calcium

Calcium is not normally low in the bloodstream because there is a large reservoir of calcium in the bones. As long as the parathyroid glands are working normally, calcium levels are maintained in the blood.

There are a few causes of low blood calcium that are not related to parathyroid hormone. The kidneys play a large role in blood calcium regulation, so kidney failure can cause calcium deficiency in the blood. You may recall that vitamin D is needed for calcium absorption. If vitamin D levels are abnormally low, then blood calcium may also be low. In cases of severe alcoholism, low magnesium levels can cause bone loss. The cells in bones that dissolve calcium out of bones, osteoclasts, become unresponsive to parathyroid hormone without enough magnesium.

Calcium Depletion from Excess Sodium

Excess dietary sodium can impact calcium levels in the body. Americans eat an average of five grams of excess sodium each day. This is the average amount above the upper intake levels. This extra sodium flushes an extra 86 mg of calcium out of the body through the kidneys each day. The most likely mechanism for this urinary loss is the competition between calcium and sodium for reabsorption in the tubules of the kidney.

If this extra 86 mg of calcium is absorbed from food sources in the intestines, then about 287 mg of extra dietary calcium is needed to account for this loss. Why is so much dietary calcium needed to replace smaller losses of urinary calcium? This is because calcium absorption from the intestines is not very efficient. Only about 30 percent of dietary calcium is absorbed from the intestines in adults (pregnant women and growing children are able to absorb about 50 percent of the calcium in food).

In some cases, the extra calcium lost in urine caused by excess sodium intake is not available from the food in the intestines. When dietary calcium is not available, this 86 mg of calcium must be removed from the bones. It is estimated that as much as one percent of the bones of adult women can be lost each year for every gram of excess sodium consumed daily. With sufficient dietary calcium, this bone loss can be prevented.

Calcium Depletion from Excess Protein

Americans regularly consume more protein than they need. On the average, women take in 24 extra grams per day and men take in an average of 44 extra grams per day of protein. This is the amount of protein consumed in excess of the RDA values of 46 grams for women and 56 grams for men. It is not uncommon for Americans to consume 100 grams of excess protein in a day.

As the intake of protein rises above the need for protein, the amount of calcium lost in urine also rises. It has been estimated that 1.75 mg of calcium is lost for each extra gram of protein consumed. Using this estimate, based upon 24 grams of excess protein, 42 mg of calcium may be lost each day because of excess protein consumption by the average American woman. With a 30 percent rate of absorption, this results in an extra dietary need of 140 mg of calcium. For the average American man with 44 grams of excess protein, the amount of extra dietary calcium needed works out to be 257 mg. If 100 grams of excess protein are eaten, the amount of dietary calcium needed to offset the protein is 583 mg. On days of heavy protein consumption, bones may need to give up calcium unless extra calcium is consumed.

> Extra calcium needed because of excess protein intake averages 140 mg for women and 257 mg for men.

The consumption of excess protein is thought to increase the amount of acids that must be neutralized by blood buffers. Protein eaten in excess of our needs is burned for energy. Calcium can be lost when it is used to neutralize the acids in blood that result from burning protein. Most of the phosphate in urine in America comes from excessive protein intakes. The phosphate forms a complex with calcium and the calcium is lost in the urine. Small amounts of calcium may also be lost as a result of the phosphoric acid in some soft drinks and the diuretic action of coffee.

Calcium and Osteoporosis

Osteoporosis is a disease that involves loss of bone strength leading to an increased risk of fracture. It is well known to occur in postmenopausal women, but can also be present in other populations. Osteoporosis afflicts 25 million people in the United States, but there are no symptoms unless a bone is broken. Hip fracture in an older person is something to be avoided as it makes movement difficult and can be slow to heal. Postmenopausal bone loss is mostly due to increases in bone loss, which are more important than decreases in bone formation.

Osteoporosis is a long-term chronic disease that normally takes decades to develop. Proper nutrition is one of the important factors in preventing osteoporosis. Supplemental calcium cannot prevent or heal osteoporosis by itself. Widely varying levels of calcium intake around the world are not associated with corresponding levels of osteoporosis. It is an unexplained paradox that osteoporosis is much more prevalent in affluent cultures with high calcium intakes than it is in poorer cultures with low calcium intakes.

High levels of calcium intake and high levels of vitamin D are helpful in slowing the progression of bone loss. Loss of calcium from the bones, as can occur on a day with lower calcium intake coupled with high sodium and protein intake, is hard to replace. Calcium can be removed quickly from bones, but it is a slower process to rebuild bones. As is true with many chronic diseases, prevention is the best approach.

One of the unavoidable risk factors is age, as bones tend to lose strength as we age. One way to decrease risk is to stay flexible and balanced, which lowers the likelihood of a fall. Weight-bearing exercise is important in early life, as it increases bone density. Exercise and movement in later life also stimulate bone formation to offset bone resorption. Some people have genetically denser bones than others, which lowers risk.

Recommended Levels of Calcium

The Food and Nutrition Board of the Institute of Medicine has set Adequate Intake levels (AI) for calcium. An RDA was not set because of the many factors that interact to affect bone health. As noted above, extra sodium and protein change the amount of calcium needed in the diet. If vitamin D is inadequate for long periods of time, lower levels of circulating calcidiol (the precursor form of vitamin D) may limit calcium absorption. Bone formation also requires vitamin K, vitamin A, magnesium, and potassium. Additionally, weight-bearing exercise helps strengthen bones.

The Food and Nutrition Board has set the AI levels high enough to compensate for extra protein, extra sodium, and low levels of vitamin D. Please refer to Table 9-1 for adequate intake levels for calcium. With lower sodium and protein levels, with regular weight-bearing exercise, and with adequate vitamin D from sun exposure, less calcium is needed to ensure bone health. Upper intake levels for calcium have been set at 2500 mg daily for ages one and above.

Food Sources of Calcium

Everyone knows that dairy products are high in calcium. Dairy products provide about 75 percent of the calcium in American diets. In spite of this, children in their highest growth years receive only 10 to 25 percent of their needed calcium. Some of the excellent sources of calcium also are excellent sources of

Table 9-1   Adequate intake levels for calcium for all ages.

Adequate Intake for Calcium	Age	Males mg/day	Females mg/day
Infants	0–6 months	210	210
Infants	7–12 months	270	270
Children	1–3 years	500	500
Children	4–8 years	800	800
Children	9–13 years	1,300	1,300
Adolescents	14–18 years	1,300	1,300
Adults	19–50 years	1,000	1,000
Adults	51 years and older	1,200	1,200
Pregnancy	18 years and younger	—	1,300
Pregnancy	19 years and older	—	1,000
Breastfeeding	18 years and younger	—	1,300
Breastfeeding	19 years and older	—	1,000

many other nutrients. Many vegetables provide abundant calcium along with high levels of vitamins A, C, K and folate; please refer to Graph 9-1. Whole sesame seeds are one of the richest sources of calcium and also provide much-needed vitamin E and selenium. Nuts, seeds, and legumes are healthy sources of calcium. Tofu is often set with calcium and can be a very high source of dietary calcium.

Calcium content of selected foods

Food	Calcium In 100 grams	Calcium In one serving
Sesame seeds	975 mg	700 mg
Cheddar cheese	704 mg	561 mg
Almonds	242 mg	139 mg
Milk, 1 cup	204 mg	504 mg
Tofu	196 mg	253 mg
Fast food hamburger	135 mg	119 mg
Soy beans	100 mg	100 mg
Kale	70 mg	187 mg

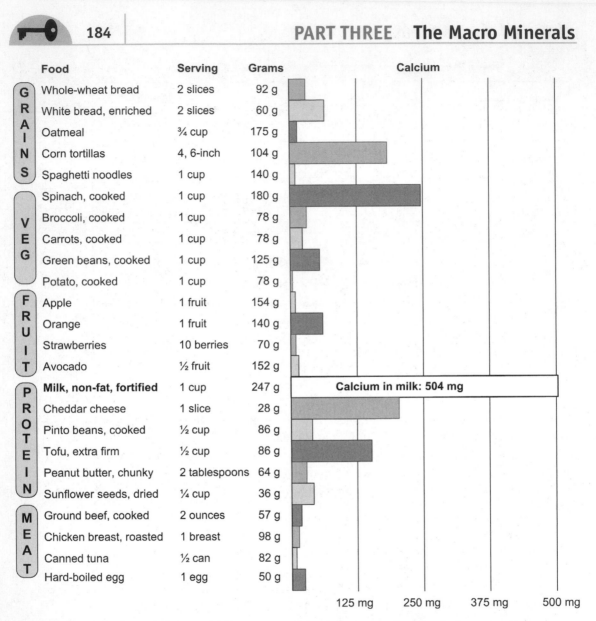

Graph 9-1   Calcium content of some common foods.

Calcium Absorption

Calcium enters the intestines from two sources. One of the sources is from the diet, which may include supplements. Calcium is also recycled back to the digestion in digestive juices. Calcium is absorbed in the upper part of the intestines in an active transport process. With higher levels of calcium intake, diffusion also plays a role in absorption from the intestines into the blood. The absorption of calcium from the intestines is controlled by the active form of vitamin D, calcitriol. Calcium absorption is highest at 400 mg daily with about 30 percent absorption. Higher intakes of calcium result in lower absorption. Calcium absorption diminishes in old age, partly because of lower calcitriol levels.

Not only is calcium absorption low, but losses are rather high. The first 400 mg of dietary calcium is needed just to cover normal losses. About 200 mg of unabsorbed calcium from both the diet and from digestive juices is lost in the feces each day. The skin, hair, and nails lose an average of 60 mg daily, necessitating an additional intake of 200 mg (calculated at 30 percent absorption). Oxalates can bind to calcium and cause excretion, but this effect is not significant in most diets.

Calcium Supplements

Calcium in supplements is found in many forms. The least absorbable forms are calcium carbonate, dolomite, and oyster shells. These forms are sometimes contaminated with lead, arsenic, cadmium, or mercury. Absorption of these forms of calcium is usually well under 10 percent. These elemental forms of calcium can easily bind to dietary oxalates and phytates, which results in their excretion. Calcium in food is normally *chelated* to other nutrients. Minerals are chelated when they are bound to organic molecules.

Calcium in supplements is often chelated to citrates, lactates, and gluconates. These forms are easily broken apart in the stomach, which limits their absorption. If taken with food, the free calcium can combine with amino acids in the food, which aids their absorption. Some supplementary calcium comes already chelated with amino acids. Amino acid chelates are not easily broken up in the stomach and are readily taken up by intestinal cells. Certain chelating agents can act as mineral transporters. Calcium aspartate, calcium ascorbate, and calcium orotate are very well absorbed and easily transported to the cells where they are needed.

Most of the calcium needed should come from the diet, rather than supplements. Calcium is a bulky nutrient and a whole day's supply will not fit conveniently into a tablet. For best absorption, iron or zinc supplements should not be

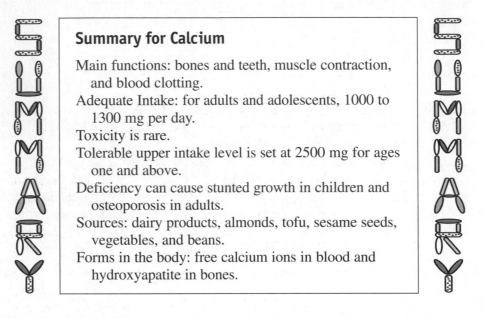
taken at the same time as calcium supplements. Tetracycline should also be taken at a different time than calcium supplements.

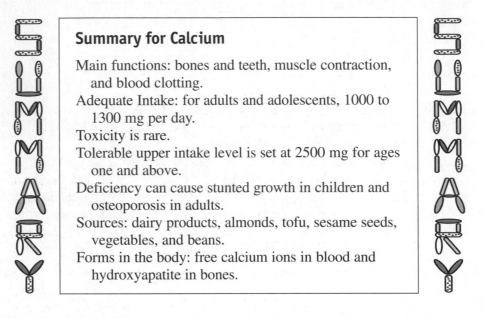

Summary for Calcium

Main functions: bones and teeth, muscle contraction, and blood clotting.

Adequate Intake: for adults and adolescents, 1000 to 1300 mg per day.

Toxicity is rare.

Tolerable upper intake level is set at 2500 mg for ages one and above.

Deficiency can cause stunted growth in children and osteoporosis in adults.

Sources: dairy products, almonds, tofu, sesame seeds, vegetables, and beans.

Forms in the body: free calcium ions in blood and hydroxyapatite in bones.

Alternatives to supplemental calcium include more calcium in the diet, more sunshine, keeping protein intake to the RDAs, keeping sodium levels below the upper intake levels, and regular weight-bearing exercise.

Calcium Toxicity

Blood levels of calcium that are abnormally high (*hypercalcemia*) are rare. Hypercalcemia has not been reported to result from food, only from certain supplement combinations. Hypercalcemia has been documented only with high levels of supplementation over long periods of time, usually combined with antacids and milk. Symptoms may include gastrointestinal disturbances, thirst, and frequent urination.

About 10 percent of Americans will experience kidney stones at one time or another. Kidney stones are normally composed of calcium oxalate and sometimes composed of calcium phosphate. Kidney stone formation has not been found to be caused by higher dietary intakes of calcium. However, kidney stones are related to the excretion of high levels of calcium in the urine. Factors that increase urinary calcium, such as excess dietary sodium and excess dietary protein, increase the risk of kidney stones. Calcium combines with oxalates in the stomach, thus removing them from absorption. The unabsorbed calcium and oxalates are passed out into the stool. By reducing the absorption of oxalates, calcium can reduce the incidence of kidney stones.

Lead toxicity can be reduced in two ways by adequate calcium intake. Less lead is absorbed from the intestines when dietary calcium levels are higher. Also, adequate dietary calcium reduces bone demineralization. Lead can be stored in the bones for decades. Higher calcium levels prevent calcium removal from the bones, and lead stored in the bones will not be mobilized into the bloodstream, either.

To summarize, calcium is famous for its use in building strong bones. Care must be taken to eat enough calcium in foods and to limit excess sodium and protein; this prevents calcium from being removed from the bones. Calcium is also needed for muscle contraction and blood clotting.

Quiz

Refer to the text in this chapter if necessary. A good score is at least 8 correct answers out of these 10 questions. The answers are listed in the back of this book.

1. The mineral component of bones, consisting of calcium and phosphate is:
    (a) Nitrogen.
    (b) Hydroxyapatite.
    (c) Oxygen.
    (d) Carbon.

2. Bones are built up with:
    (a) Resorption.
    (b) Osteoclasts.
    (c) Osteoblasts.
    (d) Osteoporosis.

3. Needed to work with the calcium channels for muscle contraction:
    (a) Troponin-c.
    (b) Calmodulin.
    (c) Activator proteins.
    (d) All of the above.

4. The quickest way for the body to raise blood calcium levels:

   (a) Absorb more calcium from the digestive tract.

   (b) Reabsorb more calcium in the kidneys.

   (c) Remove calcium from the brain.

   (d) Remove calcium from the bones.

5. Which is NOT a risk factor for low blood calcium?

   (a) Low levels of calcium in the diet.

   (b) Kidney damage.

   (c) Long-term low vitamin D intake and production.

   (d) Low magnesium levels.

6. If 100 grams of excess protein is consumed, the amount of additional dietary calcium needed is:

   (a) None.

   (b) 58 mg.

   (c) 583 mg.

   (d) 5830 mg.

7. The following is useful in slowing the progression of osteoporosis:

   (a) Extra vitamin D, preferably from sunlight.

   (b) Calcium supplementation.

   (c) Regular weight-bearing exercise.

   (d) All of the above.

8. The richest source of calcium per 100 grams is:

   (a) Sesame seeds.

   (b) Milk.

   (c) Hamburger.

   (d) Kale.

9. Absorption of dietary calcium averages:

   (a) Tenth percent.

   (b) Thirty percent.

   (c) Fifth percent.

   (d) Seventy percent.

10. Calcium supplements with the best absorption are composed of:
    (a) Calcium carbonate.
    (b) Dolomite.
    (c) Calcium ascorbate.
    (d) Oyster shell.

CHAPTER 10

Major Minerals
Phosphorus, Magnesium, and Sulfur

Phosphorus

Phosphorus is needed in every cell in the body. The name means "lightbearer," which comes from the Greek word for light, phôs, and the Greek word for bearer, phoros. The chemical symbol for phosphorus is the letter P. About 85 percent of the phosphorus in the body is in the bones. Most of the phosphorus in the body is found in the form of phosphates. Phosphates are the salts of phosphorus. Phosphates are made up of a central phosphorus atom surrounded by four oxygen atoms (PO_4), as shown in Figure 10-1.

FUNCTIONS OF PHOSPHORUS IN THE BODY

The main use for phosphorus is strengthening bones. Phosphorus and calcium form a salt called *hydroxyapatite* that provides the compressional strength of bones.

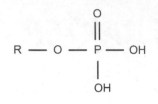

Figure 10-1   The structure of phosphate, where R is attached to the phosphate.

Phosphorus in the Cell Membrane

Another important role of phosphorus is as part of the cell membrane. All cell membranes contain *phospholipids* (*phospho* from phosphorus and *lipids* from fats). Phospholipids are major structural components of cell membranes, as seen in Figure 10-2. Phospholipids in cell membranes help control the transport of nutrients. The heads of phospholipids are attracted to water, while the tails are repelled by water. The water-loving heads of these phospholipids face towards the inside of the cell and towards the outside of the cell. The fat-loving tails face toward the inside of the cell membrane. This forms a membrane that is selectively permeable and very flexible. The cell membrane has an interior that is fat soluble while the inside and the outside of the membrane can exist in a watery environment.

Phosphorus in Energy Production

Another vital role of phosphorus is in energy production and the storage of energy in the body. The energy currency of the body, when fully charged, is known as adenosine triphosphate (ATP). Adenosine can be attached to one, two, or three phosphate groups, as shown in Figure 10-3. ATP refers to adenosine with three phosphate groups attached. When used for energy, ATP discharges to adenosine diphosphate (ADP), which has only two phosphate groups attached. ADP is able to be recharged into ATP in the cell by a process that turns the chemical bonds of energy-rich molecules such as glucose into energy. The more phosphate groups attached to adenosine, the higher the energy. AMP (*adenosine monophosphate*) has the least stored energy with only one phosphate group attached.

Phosphorus is also needed to form *creatine phosphate*. Creatine phosphate is an important energy store in skeletal muscles and the brain. It is made in the liver and is transported to the muscles and the brain. Creatine phosphate can

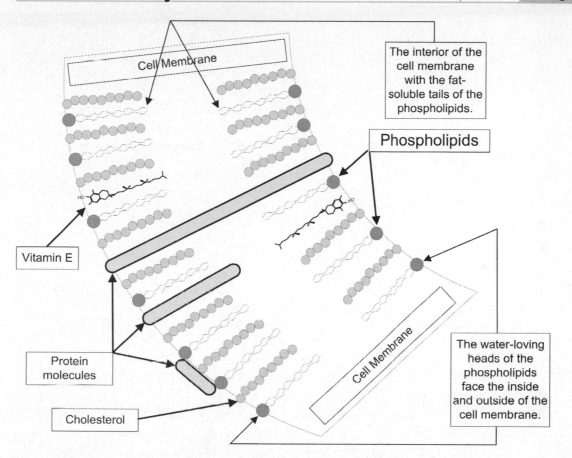

Figure 10-2   Phospholipids in the cell membrane.

pump ADP up to ATP by donating a phosphate group, as seen in Figure 10-4. This happens in just two to seven seconds after an intense effort. After donating a phosphate group, creatine phosphate becomes creatine. This creatine can be recharged back into creatine phosphate.

Phosphorus is the Backbone of DNA

Nucleic acids store and transmit genetic information. The two types of nucleic acids are DNA and RNA. They are both made of long chains of phosphate-containing molecules. Phosphates are needed for growth as they form the backbone of both DNA and RNA.

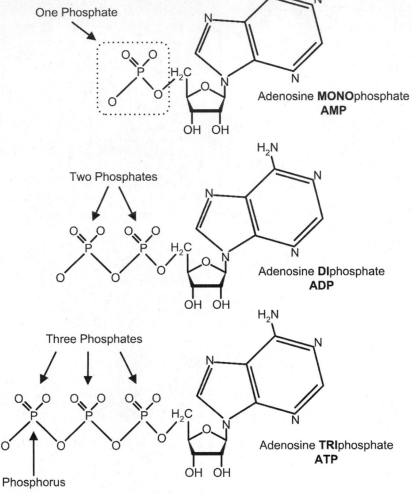

One Phosphate

Adenosine **MONO**phosphate
AMP

Two Phosphates

Adenosine **DI**phosphate
ADP

Three Phosphates

Phosphorus

Adenosine **TRI**phosphate
ATP

Figure 10-3   Phosphates and ATP.

Phosphorylation

Phosphorylation is the addition of a phosphate (PO_4) group to a molecule, usually a protein. Many enzymes and receptors are switched "on" or "off" by phosphorylation and *dephosphorylation*. The addition of a phosphate to a protein can change a protein from hydrophobic ("water-hating") to hydrophilic ("water-loving"). One

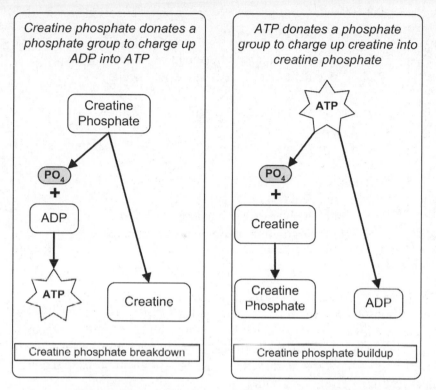

Figure 10-4 Phosphates and muscle contraction.

use of phosphorylation is when certain light-sensitive cells of the retina use phosphorylation for signaling the presence of light.

Phosphorus as a Blood Buffer

Phosphorus also helps to maintain normal acid-alkaline balance (pH) in its role as one of the body's most important buffer systems. *Dihydrogen phosphate* absorbs acids produced by metabolic activity and becomes phosphoric acid.

REGULATION OF PHOSPHORUS

Phosphorus in food is easily absorbed in the small intestine. Excesses of phosphorus are eliminated by the kidneys. Phosphorus is regulated along with calcium.

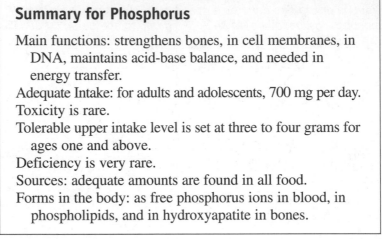

Summary for Phosphorus

Main functions: strengthens bones, in cell membranes, in DNA, maintains acid-base balance, and needed in energy transfer.

Adequate Intake: for adults and adolescents, 700 mg per day.

Toxicity is rare.

Tolerable upper intake level is set at three to four grams for ages one and above.

Deficiency is very rare.

Sources: adequate amounts are found in all food.

Forms in the body: as free phosphorus ions in blood, in phospholipids, and in hydroxyapatite in bones.

When the parathyroid glands sense low calcium levels in the blood, more calcium and phosphorus are absorbed into the blood from the intestines. If calcium levels are still low, both calcium and phosphorus are removed from the bones and put into the blood. However, while the parathyroid hormone causes more calcium to be retained by the kidneys, it increases the loss of phosphorus in the urine. This results in a higher ratio of calcium to phosphorus in the blood.

DEFICIENCY OF PHOSPHORUS

Deficiency of phosphorus in the blood is called *hypophosphatemia* (*hypo-* for low, *phosphate* for phosphorus, and *-emia* for blood). Hypophosphatemia is very rare because phosphorus is so widespread in food. In fact, low levels of phosphorus are rarely seen except in cases of near starvation, alcoholism, and certain types of diabetes.

High doses of aluminum-containing antacids can cause loss of phosphorus, and, over a period of time, low blood levels of phosphorus.

RECOMMENDED LEVELS FOR PHOSPHORUS

RDAs have been set for phosphorus to meet the needs of bones and for cellular needs. Adequate intake levels (AI) have been estimated for infants. Please refer to Table 10-1 for adequate levels of phosphorus for all ages.

Table 10-1   RDAs and adequate intakes (AI) for phosphorus for all ages.

RDAs for Phosphorus	Age	Males mg/day	Females mg/day
Infants	0–6 months	100 (AI)	100 (AI)
Infants	7–12 months	275 (AI)	275 (AI)
Children	1–3 years	460	460
Children	4–8 years	500	500
Adolescents	14–18 years	1,250	1,250
Adults	19 years and older	700	700
Pregnancy	18 years and younger	—	1,250
Pregnancy	19 years and older	—	700
Breastfeeding	18 years and younger	—	1,250
Breastfeeding	19 years and older	—	700

FOOD SOURCES OF PHOSPHORUS

Phosphorus is found in virtually all food; please refer to Graph 10-1. Americans are estimated to receive more than enough phosphorus in the diet. Phosphorus is found in the form of phosphoric acid in many soft drinks. Phosphorus levels in the diet have increased because of food additives.

The phosphorus in the seeds of plants is found in the form of *phytates* (also called *phytic acid*). Phytates lessen the availability of calcium, magnesium, and iron. In phytate or phytic acid form, only about half of the phosphorus is available for absorption. This presents little difficulty as phosphorus is present in virtually all diets in adequate amounts.

TOXICITY OF PHOSPHORUS

Excess phosphorus in the blood is called *hyperphosphatemia*. The kidneys are very efficient at eliminating excess phosphorus. Excess blood phosphorus is mainly found in people with damaged kidneys. Kidneys can become further damaged from kidney stones made from calcium phosphate deposits once kidney function is reduced to a level below 25 percent. Tolerable upper intake levels (UL) for healthy people have been set at three to four grams of phosphorus.

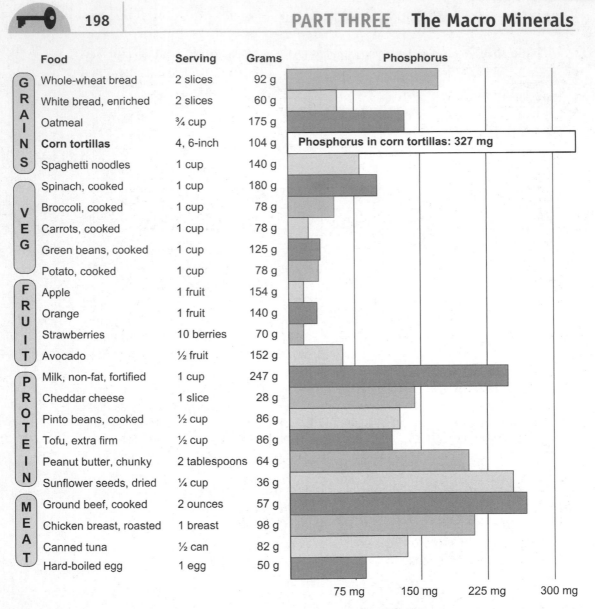

Food	Serving	Grams	Phosphorus
GRAINS			
Whole-wheat bread	2 slices	92 g	
White bread, enriched	2 slices	60 g	
Oatmeal	¾ cup	175 g	
Corn tortillas	4, 6-inch	104 g	Phosphorus in corn tortillas: 327 mg
Spaghetti noodles	1 cup	140 g	
VEG			
Spinach, cooked	1 cup	180 g	
Broccoli, cooked	1 cup	78 g	
Carrots, cooked	1 cup	78 g	
Green beans, cooked	1 cup	125 g	
Potato, cooked	1 cup	78 g	
FRUIT			
Apple	1 fruit	154 g	
Orange	1 fruit	140 g	
Strawberries	10 berries	70 g	
Avocado	½ fruit	152 g	
PROTEIN			
Milk, non-fat, fortified	1 cup	247 g	
Cheddar cheese	1 slice	28 g	
Pinto beans, cooked	½ cup	86 g	
Tofu, extra firm	½ cup	86 g	
Peanut butter, chunky	2 tablespoons	64 g	
Sunflower seeds, dried	¼ cup	36 g	
MEAT			
Ground beef, cooked	2 ounces	57 g	
Chicken breast, roasted	1 breast	98 g	
Canned tuna	½ can	82 g	
Hard-boiled egg	1 egg	50 g	

75 mg   150 mg   225 mg   300 mg

Graph 10-1 Phosphorus content of some common foods.

To summarize, phosphorus is a versatile nutrient. It strengthens bones and cell membranes. Phosphorus is used to store energy in the body and to maintain the acid-base balance.

Magnesium

Magnesium is essential for more than 300 metabolic reactions in cells in concert with hundreds of enzyme systems. The chemical symbol for magnesium is *Mg*. Magnesium assists with the synthesis of nucleic acids, fats, and protein. Magnesium offsets calcium for properly controlled blood clotting. Magnesium holds calcium in teeth to make teeth more resistant to cavities. Although there is less than one ounce (about 19 grams) of magnesium in an adult body, it is a vital nutrient. Most of the magnesium in the body is found in the skeleton, as seen in Figure 10-5. About one quarter of the magnesium is found in muscles. Only one percent of the magnesium in the body is found outside of cells. Magnesium in blood cells has three forms: bound to proteins, in stable compounds, and as ionized magnesium (Mg^{++}).

Aerobic energy production in the cell requires magnesium. The enzyme that controls the first four steps of the oxygen energy cycle requires magnesium for its activity. Deficiency of magnesium can slow down this energy-producing cycle.

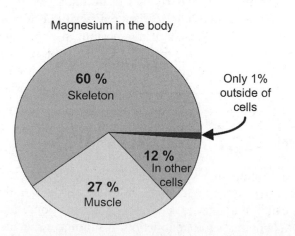

Figure 10-5 Magnesium is normally found inside cells.

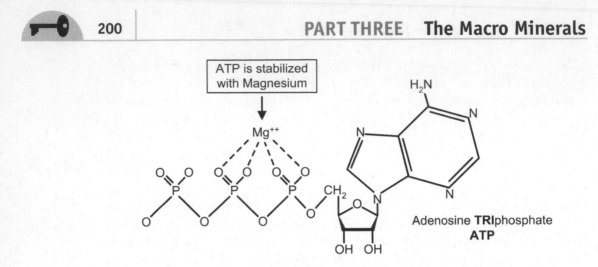

Figure 10-6   ATP is stabilized with magnesium.

Most energy production and energy transfer in cells use ATP, which exists as a complex with magnesium. When ATP is synthesized, magnesium is used to stabilize the phosphates, as seen in Figure 10-6.

MAGNESIUM AND BONE MINERALIZATION

Magnesium is part of the structure of bones. Most of the magnesium in the body, about 60 percent, is used as part of the structure of the bones. Magnesium is needed for proper bone mineralization along with calcium and phosphorus. Parathyroid hormone and calcitriol (the active vitamin D hormone) both depend upon magnesium for the mineralization of bone. When magnesium is low in the blood, blood calcium levels also fall. Low levels of magnesium in bone cause a resistance to parathyroid hormone, resulting in less bone mineralization. Even lower levels of magnesium in bone lead to bone crystals that are larger and more brittle. This is why adequate magnesium may be a factor in preventing osteoporosis.

MUSCLE CONTRACTION AND RELAXATION

Magnesium and calcium work together to coordinate muscle contraction and relaxation. These two minerals maintain normal cell membrane electrical potentials. Increased calcium in muscle cells triggers contraction. Increased magnesium in muscle cells counteracts the contraction, resulting in relaxation. Calcium and magnesium also work together to contract and relax capillaries. Magnesium is a potent vasodilator, opening blood vessels and lowering blood pressure.

MAGNESIUM DEFICIENCY

Serious deficiencies of magnesium are rare because magnesium is common in most food. Also, the kidneys limit losses of magnesium when intake is low. Nevertheless, average intake levels of magnesium in America are below the recommended levels. On the average, Americans eat about three-quarters of the RDA of magnesium. In older Americans intakes of magnesium are even lower.

Certain digestive disturbances can limit magnesium absorption. Irritable bowel syndrome and prolonged diarrhea have been associated with magnesium depletion. Kidney disorders may cause loss of magnesium. Magnesium depletion is frequently encountered in chronic alcoholics, both from low dietary intake and increased urinary losses. Some alcohol withdrawal symptoms may be related to magnesium deficiency. In elderly populations absorption of magnesium may be lower. Also, magnesium losses in urine increase in older people. These factors, coupled with lower intakes, increase the risk of magnesium depletion in the elderly.

Summary for Magnesium

Main functions: strengthens bones, promotes muscle
    relaxation, and used to stabilize ATP.
Adequate Intake: for adults, 300 mg to 420 mg per day.
Toxicity is possible from supplement use and may result in
    diarrhea.
Tolerable upper intake level is set at 350 mg for supple-
    mental doses.
Deficiency is very rare. Intakes below the RDAs are common.
Sources: whole grains, nuts, leafy greens, and seeds.
Forms in the body: in bones, bound to protein, attached
    to ATP, and as ionized magnesium (Mg^{++}).

Low levels of magnesium in the blood lead to low levels of calcium and potassium in the blood. Low blood levels of magnesium (*hypomagnesemia*) can cause muscular symptoms such as spasms and muscle tremors. Other symptoms may include digestive problems and personality changes.

SOURCES OF MAGNESIUM

Green leafy vegetables are a good source of magnesium because magnesium is the central ion in chlorophyll. Whole grains, nuts, and seeds are rich sources of magnesium; please refer to Graph 10-2. Unfortunately, when whole wheat is refined

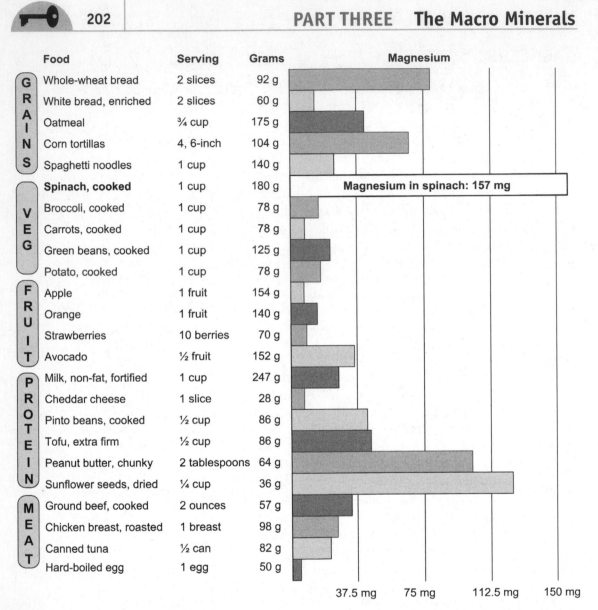

Food	Serving	Grams	Magnesium
GRAINS			
Whole-wheat bread	2 slices	92 g	
White bread, enriched	2 slices	60 g	
Oatmeal	¾ cup	175 g	
Corn tortillas	4, 6-inch	104 g	
Spaghetti noodles	1 cup	140 g	
VEG			
Spinach, cooked	1 cup	180 g	Magnesium in spinach: 157 mg
Broccoli, cooked	1 cup	78 g	
Carrots, cooked	1 cup	78 g	
Green beans, cooked	1 cup	125 g	
Potato, cooked	1 cup	78 g	
FRUIT			
Apple	1 fruit	154 g	
Orange	1 fruit	140 g	
Strawberries	10 berries	70 g	
Avocado	½ fruit	152 g	
PROTEIN			
Milk, non-fat, fortified	1 cup	247 g	
Cheddar cheese	1 slice	28 g	
Pinto beans, cooked	½ cup	86 g	
Tofu, extra firm	½ cup	86 g	
Peanut butter, chunky	2 tablespoons	64 g	
Sunflower seeds, dried	¼ cup	36 g	
MEAT			
Ground beef, cooked	2 ounces	57 g	
Chicken breast, roasted	1 breast	98 g	
Canned tuna	½ can	82 g	
Hard-boiled egg	1 egg	50 g	

37.5 mg   75 mg   112.5 mg   150 mg

Graph 10-2   Magnesium content of some common foods.

into enriched white flour, the magnesium for a 100-gram serving is reduced from 126 mg to 22 mg. Processed and refined foods are generally low in magnesium. Fast foods are not an optimal source; it would take seventeen fast-food double hamburgers to provide enough magnesium for one day. Chocolate is a fun source of magnesium.

The RDAs have been set for magnesium to prevent deficiency. Adequate intakes (AI) have been estimated for infants, as seen in Table 10-2.

About 40 to 50 percent of dietary magnesium is absorbed. With small amounts of dietary magnesium, absorption is higher. With higher intakes, absorption is lower. Most dietary magnesium is absorbed in the colon or far down in the intestines.

Magnesium supplements are available in poorly absorbed forms such as magnesium oxide and magnesium chloride. The poor solubility of these salts diminishes their absorption. Absorption is a little better with magnesium gluconate and magnesium citrate. Amino acid chelates such as magnesium aspartate are better absorbed. Magnesium orotate and magnesium ascorbate are well absorbed and transport of magnesium to the cell is facilitated.

Table 10-2 RDAs and adequate intakes (AI) for magnesium for all ages.

RDAs for Magnesium	Age	Males mg/day	Females mg/day
Infants	0–6 months	30 (AI)	30 (AI)
Infants	7–12 months	75 (AI)	75 (AI)
Children	1–3 years	80	80
Children	4–8 years	130	130
Children	9–13 years	240	240
Adolescents	14–18 years	410	360
Adults	19–30 years	400	310
Adults	31 years and older	420	320
Pregnancy	14–18 years	—	400
Pregnancy	19–30 years	—	350
Pregnancy	31–50 years	—	360
Breastfeeding	14–18 years	—	360
Breastfeeding	19–30 years	—	310
Breastfeeding	31–50 years	—	320

TOXICITY OF MAGNESIUM

Magnesium in food has not been associated with any side effects. The tolerable upper intake level has been set at 350 mg for supplemental magnesium in adults and adolescents. This level is set to minimize diarrhea, which is a common side effect of high supplemental magnesium intakes. In fact, larger amounts of magnesium are used in laxative formulas. People with kidney impairment may need to lower any supplemental dosage below the tolerable upper intake levels. High levels of magnesium in the blood can result in lowered blood pressure and lethargy.

To summarize, magnesium is found in the center of chlorophyll and is needed to relax muscles and to strengthen bones.

Sulfur

There is more sulfur in the human body than magnesium, sodium, or chloride. The chemical symbol for sulfur is the letter *S*. Sulfur is a major mineral that often occurs with other nutrients. Sulfur is part of the important antioxidant *glutathione*. Sulfur is also an important part of *coenzyme A*, which is central to energy metabolism. Sulfur is part of the methyl donor S-adenosylmethionine (SAMe). Sulfur is part of biotin and vitamin B_1 (thiamin).

> ### Summary for Sulfur
>
> Main functions: needed for the antioxidant glutathione, part of coenzyme A, and part of the methyl donor SAMe.
> Adequate Intake: none established.
> Toxicity is not known.
> Tolerable upper intake level has not been set.
> Deficiency: not known.
> Sources: food with sulfur-containing amino acids (methionine and cysteine), onions, garlic, cabbage, and Brussels sprouts.
> Forms in the body: found in glutathione, coenzyme A, methionine, cysteine, and SAMe.

The two sulfur-containing amino acids are methionine and cysteine. Protein can be stabilized with the amino acid cysteine. Cysteine has a side chain that contains sulfur. These side chains can link together to form *disulfide bridges* (disulfide means

Human Insulin

Disulfide bridges can form between two cysteine (Cys) amino acids because they have side groups containing sulfur. The disulfide bridges stabilize the protein insulin. The two polypeptide chains are made up of amino acids.

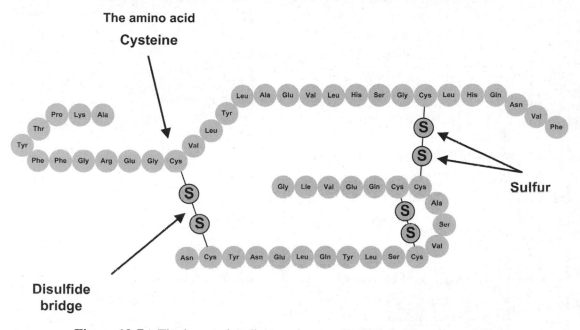

Figure 10-7   The human insulin protein uses disulfide bridges for stability.

two sulfurs). These disulfide bridges stabilize protein structures, as shown with insulin in Figure 10-7. The disulfide bridges also help to fold and shape proteins.

The more rigid tissues such as those found in skin, hair, and nails have a high content of the protein keratin. Keratin is a type of protein that is tough and insoluble. Disulfide bridges give extra strength to nails and hair by cross-linking the keratin. The more cross-linking, the stronger the tissue becomes. Human hair is approximately 14 percent cysteine. The pungent smell of burning hair is due to the sulfur compounds in hair.

Deficiencies of sulfur are not known except in cases of severe protein deficiency. There are no recommended intakes for sulfur and there are no upper levels set. Sulfur is easily and efficiently absorbed as part of cysteine and methionine or as inorganic sulfate.

To summarize, sulfur is vital for antioxidant activity and energy production. Sulfur is also used to fold and stabilize some proteins. Sulfur is not as well known as many other essential nutrients.

Quiz

Refer to the text in this chapter if necessary. A good score is at least 8 correct answers out of these 10 questions. The answers are listed in the back of this book.

1. Phospholipids:
    (a) Are made of fat and phosphorus.
    (b) Are a major constituent of the cell membrane.
    (c) Control the transport of nutrients into the cell.
    (d) All of the above.

2. Adenosine triphosphate is made up of:
    (a) One phosphate group and adenosine.
    (b) Two phosphate groups and adenosine.
    (c) Three phosphate groups and adenosine.
    (d) Four phosphate groups and adenosine.

3. Most of the magnesium in the body is found in the
    (a) Skeleton.
    (b) Blood.
    (c) Brain.
    (d) Kidneys.

4. Magnesium is used to stabilize:
    (a) Blood acidity.
    (b) The phosphate groups on ATP.
    (c) Psychiatric patients.
    (d) Blood sugar.

5. Which mineral is NOT used in the structure of bones?
    (a) Phosphorus.
    (b) Calcium.
    (c) Sulfur.
    (d) Magnesium.

6. Increased magnesium in muscle cells results in:
    (a) Increased relaxation.
    (b) Increased contraction.
    (c) Increased muscle tone.
    (d) Faster contraction.

7. Average amounts of magnesium intake for Americans is:
    (a) Less than the RDAs.
    (b) The same as the RDAs.
    (c) Higher than the RDAs.
    (d) Almost none.

8. A good source of magnesium:
    (a) Hamburgers.
    (b) Enriched bread.
    (c) Nuts and seeds.
    (d) All of the above.

9. A sulfur-containing amino acid:
    (a) Arginine.
    (b) Cysteine.
    (c) Tryptophan.
    (d) Lysine.

10. Insulin and other proteins can be stabilized with:
    (a) One sulfur ion.
    (b) Monosulfide bridges.
    (c) Trisulfide bridges.
    (d) Disulfide bridges.

Test: Part Three

Do not refer to the text when taking this test. A good score is at least 18 (out of 25 questions) correct. Answers are in the back of the book. It's best to have a friend check your score the first time, so that you won't memorize the answers if you want to take the test again.

1. Water is needed:
   (a) Less often than vitamins.
   (b) Less often than minerals.
   (c) More often than vitamins or minerals.
   (d) Less often than vitamins or minerals.

2. The body loses the most amount of water through:
   (a) The lungs.
   (b) Skin diffusion.
   (c) The feces.
   (d) The kidneys.

3. The kidneys retain more water:

   (a) When blood pressure drops.

   (b) In response to antidiuretic hormone.

   (c) In response to aldosterone hormone.

   (d) All of the above.

4. Which food type has the most water?

   (a) Broccoli.

   (b) Bread.

   (c) Cheese.

   (d) Crackers.

5. Which is NOT an electrolyte mineral?

   (a) Potassium.

   (b) Iron.

   (c) Sodium.

   (d) Chloride.

6. Which fluid comprises about 40 percent of weight?

   (a) Intracellular fluid.

   (b) Interstitial fluid.

   (c) Extracellular fluid.

   (d) Blood plasma.

7. Ions are charged particles. Which are positively charged?

   (a) Phosphate.

   (b) Electrons.

   (c) Cations.

   (d) Anions.

8. Outside the cell, which are the most common electrolytes?

   (a) Magnesium and phosphate.

   (b) Sodium and chloride.

   (c) Sodium and magnesium.

   (d) Chloride and potassium.

9. The most important organ for maintaining water balance is:

(a) Lungs.

(b) Skin.

(c) Liver.

(d) Kidneys.

10. The acid-alkaline balance:

(a) Keeps blood at 7.35 to 7.45 pH.

(b) Regulates the acidity of urine.

(c) Pumps sodium out of the cells.

(d) Removes water from the body.

11. This amount of carbonic acid is released by the lungs as carbon dioxide daily:

(a) Five liters (about 1 gallon).

(b) Ten liters (about 2.5 gallons).

(c) Twenty liters (about 5 gallons).

(d) Thirty liters (about 8 gallons).

12. The percentage of salt added at the table and in the kitchen is:

(a) 10 percent.

(b) 20 percent.

(c) 30 percent.

(d) 40 percent.

13. Adequate potassium protects against:

(a) Lung problems.

(b) Acidosis that can increase the risk of osteoporosis.

(c) Kidney failure.

(d) Hair loss.

14. Processed food has:

(a) Less potassium and more sodium.

(b) Less sodium and more potassium.

(c) More potassium and less chloride.

(d) Less sodium and chloride.

15. During peak growth years, the amount of extra calcium retained each day is:
    (a) 100 mg.
    (b) 200 mg.
    (c) 300 mg.
    (d) 400 mg.

16. Bones are dissolved by:
    (a) Magnesium.
    (b) Osteoclasts.
    (c) Osteoblasts.
    (d) Osteoporosis.

17. Calcium ions in muscle cells can:
    (a) Stimulate muscle contraction.
    (b) Release blood sugar from storage.
    (c) Enter through the calcium channel.
    (d) All of the above.

18. Kidneys release which substance to raise blood calcium levels?
    (a) Calcitriol.
    (b) Calcidiol.
    (c) Cholecalciferol.
    (d) Cholesterol.

19. Calcium can be depleted from the bones by:
    (a) Excess vitamin D.
    (b) Excess sodium and protein.
    (c) Excess vitamin A.
    (d) Excess vitamin E.

20. Calcium removed from bones is:
    (a) Replaced quickly.
    (b) Impossible to replace.
    (c) Time-consuming to replace.
    (d) Never removed from bones.

21. Which type of calcium supplement is best absorbed?
    (a) Calcium carbonate.
    (b) Dolomite.
    (c) Calcium gluconate.
    (d) Calcium ascorbate.

22. Lead toxicity is:
    (a) Reduced by adequate calcium intake.
    (b) Increased by adequate calcium intake.
    (c) Unaffected by calcium intake.
    (d) Increased by calcium citrate supplements.

23. Phosphorus in cell membranes is in the form of:
    (a) Phosphate.
    (b) Phosphorus ions.
    (c) Phospholipids.
    (d) Phosphoproteins.

24. Dihydrogen phosphate buffers blood acids and becomes:
    (a) Sulfuric acid.
    (b) Phosphoric acid.
    (c) Carbonic acid.
    (d) All of the above.

25. The phosphate groups in ATP are stabilized with:
    (a) Sulfur.
    (b) Phosphorus.
    (c) Calcium.
    (d) Magnesium.

PART FOUR

The Trace Minerals

Introduction to the Trace Minerals

In addition to the major minerals, the human body takes in many of the other elements in the periodic table. Many of these minerals are known to be essential in the diet for life and health. As time passes, more of these minerals are being found to have important roles. Deficiencies of two of the most important trace minerals, iron and iodine, are responsible for widespread disease. Certain trace minerals, such as lead and mercury, are most noted for their toxicity.

Food processing and cooking can remove significant quantities of trace minerals. The production of white flour from whole wheat results in large losses of many trace minerals. Although iron is added back to the white flour, many of the other trace minerals are not. Cooking, especially boiling, causes leaching of trace minerals into water that may be thrown out. Zinc, copper, and chromium are often found in low amounts in typical American diets.

Absorption of dietary trace minerals varies considerably. Only about one percent of dietary chromium is absorbed and only about 10 percent of dietary iron is absorbed. In contrast, almost all of dietary cobalt, molybdenum, and iodine can be absorbed. Many of the metallic trace minerals are bound to *transferrin* for transport in blood plasma. Iron, manganese, and chromium use transferrin for transport. *Albumin*, another plasma protein, is used to transport fluoride, zinc, and cobalt. Excesses of most of these trace minerals are eliminated by the kidneys. However, iron, zinc, copper, and molybdenum are eliminated in the bile.

Perfect health is not possible without the correct amounts of these trace minerals.

CHAPTER 11

Iron
The Blood Builder

Although iron is necessary for our bodies, it can be harmful in large amounts. The chemical symbol for iron is *Fe* from the Latin *ferrum*, which means iron. Iron deficiency is probably the most common nutrient deficiency in the United States and the world. Iron deficiency affects about one billion people worldwide.

Iron is used to transport and store oxygen in blood and to store oxygen in muscles. Iron is useful because it has the ability both to oxidize and to *reduce* (the opposite of oxidize). Oxidation describes the loss of an electron. Just the opposite, reduction is the gaining of an electron. One of the uses of iron in the body is in the electron transport chain, which is the principal source of energy in cells. Iron transports electrons by accepting and releasing them in this electron transport chain.

Iron in the cells is normally bound to proteins. If iron is found free within the cell, it can cause oxidation and free radical damage.

Iron has several vitally important roles in the body. Iron is needed for the synthesis of DNA. This synthesis of DNA is vital to support growth, healing, reproduction, and immune function. Iron is also required by enzymes involved in synthesizing collagen, neurotransmitters, and hormones.

Iron Transports and Stores Oxygen

Hemoglobin and *myoglobin* are proteins that transport and store oxygen. *Heme* is a compound found in both of these proteins. Hemoglobin is composed of four units; each unit has a protein chain and a heme group. Myoglobin has one protein chain and one heme group. Heme is a compound that has iron in the center, as seen in Figure 11-1.

Hemoglobin is the most important protein found in red blood cells. About two-thirds of the iron in the body (about 2.5 grams) is inside the heme found in hemoglobin, as seen in Figure 11-2. Hemoglobin has the unique ability to very rapidly acquire oxygen from the lungs. Hemoglobin then transports oxygen to the rest of the body. Hemoglobin releases oxygen as needed as the blood moves through the tissues.

Myoglobin (*myo* means muscle) is a protein that contains heme and is found in muscle cells. About 15 percent of the iron in the body is found in myoglobin. Myoglobin has a great affinity for oxygen, which it binds and holds. The oxygen

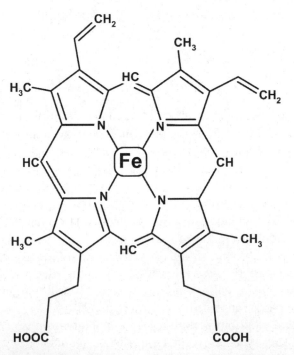

Figure 11-1   Heme in red blood cells with iron (Fe) in the center.

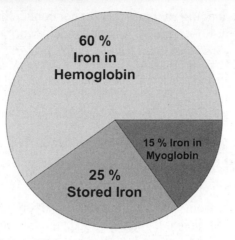

Figure 11-2   Iron distribution in the body.

is bound directly to the iron atom inside myoglobin. The myoglobin then stores the oxygen inside muscle cells. When oxygen is not readily available from blood, myoglobin releases its stored oxygen for the use of the muscle cell.

Iron in Energy Production

Iron is important to energy production in the cell in several different ways. Iron-containing heme is an essential component of *cytochromes*. Cytochromes are important for aerobic energy production as part of the electron transport chain. They serve as electron carriers during the synthesis of ATP (adenosine triphosphate), the primary energy storage compound in cells. In the electron transport chain, hydrogen (H) and oxygen (O_2) are combined into water (H_2O) plus electrons. The iron-containing cytochromes pass the electrons along the transport chain.

An iron-sulfur protein is also used in the electron transport chain, which pumps up ATP. The final step in the electron transport chain uses a complex containing two copper atoms and two compounds of heme.

An enzyme that contains iron and transports electrons is *succinate dehydrogenase*. This enzyme contains heme and is very important in energy metabolism. Succinate dehydrogenase also contains riboflavin (vitamin B_2). This enzyme functions at the crossroads between the tricarboxylic acid (TCA) cycle and the electron transport chain.

There are iron-containing enzymes that do not contain heme that are important to energy metabolism. For example, iron and niacin work together in the enzyme *NADH (nicotinamide adenine dinucleotide hydrogenase)* to transport electrons in the electron transport chain. You may recall that *nicotinamide* is a form of niacin, vitamin B$_3$.

Iron as an Antioxidant

Cells need protection from any accumulation of hydrogen peroxide. Hydrogen peroxide can cause free radical damage inside cells. Certain enzymes that contain heme can catalyze reactions that neutralize the free radical *hydrogen peroxide*. Hydrogen peroxide is converted to water and oxygen, as shown in Figure 11-3. In this case, iron in heme acts as an antioxidant.

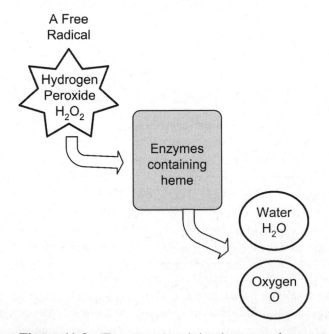

Figure 11-3 Enzymes containing heme catalyze the conversion of hydrogen peroxide to water and oxygen.

Iron and the Immune System

Iron is needed by the immune system. For instance, iron is needed for the growth of T lymphocytes. Also, the body uses the free radical action of iron to attack some pathogens. Iron helps the immune system in another way by enhancing the ability of white blood cells to engulf and kill bacteria. In this case, bacteria are attacked by free radicals that are created using an iron catalyst.

> **Bacteria Need Iron**
>
> The body lowers blood iron in response to infections.

Iron is also needed by infectious agents such as bacteria. The body lowers blood iron levels during acute infections to avoid aiding the infectious agents. Lowering blood iron may be an important immune response to infections. Iron fortification and supplementation must be reconsidered during infections to prevent worsening infections such as malaria, tuberculosis, HIV (human immunodeficiency virus), and typhoid.

Nutrient Interactions with Iron

Iron deficiency anemia may be aggravated by a deficiency of vitamin A. With iron deficiency anemia, the combination of iron supplementation with vitamin A supplementation is more effective than either supplement alone. Vitamin A seems to improve the iron status of pregnant women and children.

Copper has been found to assist iron absorption. Copper may assist the transport of iron from the liver to the bone marrow for red blood cell formation. Anemia is one of the clinical signs of both copper deficiency and iron deficiency.

Iron supplementation has been found to interfere with the absorption of zinc supplements, but only on an empty stomach. Taken with food, iron does not interfere with zinc absorption. Most supplements are designed to be taken with food and this is one reason why. The iron in fortified food does not appear to inhibit zinc absorption.

Calcium can interfere with iron absorption when both nutrients are consumed together. The calcium and phosphorus in milk can reduce iron absorption. This is not normally a problem, as there does not seem to be any interference when the amount of calcium is under 1000 mg (the amount in two cups of milk).

> **Nutrient Interactions with Iron**
>
> Vitamin A helps relieve iron deficiency.
> Copper assists iron absorption.
> Iron can interfere with zinc absorption on
>   an empty stomach.
> High levels of calcium supplementation
>   can reduce iron absorption.

Iron Deficiency

The highest prevalence of iron deficiency is found in infants, children, adolescents, and women of childbearing age, especially pregnant women. Infants have very high iron requirements during the weaning period.

Deficiency of iron in the diet does not immediately cause *iron deficiency anemia*. Stores of iron are normally depleted slowly with no anemic effects until iron stores are very low. Even before signs of anemia occur, signs of apathy and fatigue may be apparent. Some children may be incorrectly diagnosed with attention deficit disorders when they are actually suffering from iron depletion.

After iron stores are depleted, blood cells begin to have less hemoglobin and the blood cells start to become smaller than usual. With less hemoglobin than needed, oxygen delivery to the cells becomes inadequate, especially during exertion. There are other causes of anemia, such as deficiency of vitamin B_{12} or folate.

Symptoms of iron deficiency anemia are usually a result of inadequate oxygen delivery. Symptoms include fatigue, rapid heart rate, reduced work capacity, and rapid breathing upon exertion. Iron deficiency can also limit the ability to maintain body temperature in cold conditions. Both hemoglobin in blood and myoglobin in muscles become depleted. Lack of iron may also limit the creation of energy in the electron transport chain. This may lead to more anaerobic energy production resulting in excess lactic acid and fatigue. Severe iron deficiency anemia can result in spoon-shaped, brittle nails, taste bud atrophy, and mouth sores.

> Symptoms of iron deficiency anemia are usually
> a result of inadequate oxygen delivery.

Lack of sufficient iron in early childhood can contribute to problems with learning, memory, and behavior. Supplementation may prevent further problems, but may not correct existing iron deficiency problems. Iron deficiency has been found to increase the intestinal absorption of lead, which contributes to learning and memory problems.

High-Risk Individuals for Iron Deficiency

Rapid growth rates coupled with limited intakes of iron can cause deficiency in children between the ages of six months and four years. Infants are normally born with enough iron stores to last the first six months. The period of rapid growth in early adolescence is also a time of risk for iron deficiency. Adolescent girls who are growing rapidly and also menstruating are at a high risk of iron deficiency.

Pregnant women need only a normal amount of iron in early pregnancy. However, during the last trimester of pregnancy, iron needs exceed the iron present in even the best diet. During the last trimester, iron is removed from the mother's iron stores to supply the growing fetus. There is normally a deficit of about 450 mg of iron during pregnancy—about 80 percent of this deficit takes place during the last six weeks of pregnancy. Iron stores of this size are not common, so it is recommended by the World Health Organization that pregnant women take iron supplements along with folic acid supplements. It is ideal if a woman rebuilds her iron stores back up to about 500 mg before a subsequent pregnancy.

Populations with Increased Risk of Iron Deficiency

Infants under one year of age drinking cow's milk
Children aged six months to four years
Adolescent girls
Pregnant women in the last trimester
Those with blood loss including blood donors
Athletes

Loss of large amounts of blood can result in iron deficiency anemia. About one half-quart of blood (500 ml) contains about 242 mg of iron. Blood bank donations of 500 ml should be carefully considered in populations considered at risk for iron deficiency anemia. Long periods of time can be required to replace the iron in large amounts of lost blood. People with chronic losses of blood can also be at risk for iron deficiency. Hookworm infections can cause chronic losses of blood that contribute to an increased risk of iron deficiency. Menstrual blood loss can also increase risk of iron deficiency if iron loss exceeds absorbed iron.

Athletes may be at risk for iron deficiency. The average requirement for iron may be approximately 30 to 70 percent higher for those who engage in regular intense exercise.

Infants under one year of age may be at increased risk of iron deficiency if they drink cow's milk. Cow's milk consumed at this age has been associated with small losses of blood in the stool. The Food and Nutrition Board of the National Institute of Medicine recommends monitoring infants under one year of age for anemia if they drink cow's milk.

The Recommended Daily Allowance for Iron

The Food and Nutrition Board has set RDAs to prevent iron deficiency. Adequate Intake levels (AI) are set for younger infants, as seen in Table 11-1. Strict

Table 11-1   RDAs and adequate intakes (AI) for iron for all ages.

RDAs for Iron	Age	Males mg/day	Females mg/day
Infants	0–6 months	0.27 (AI)	0.27 (AI)
Infants	7–12 months	11	11
Children	1–3 years	7	7
Children	4–8 years	10	10
Children	9–13 years	8	8
Adolescents	14–18 years	11	15 (vegetarian 26)
Adults	19–50 years	8 (vegetarian 14)	18 (vegetarian 33)
Adults	51 years and older	8 (vegetarian 14)	8 (vegetarian 14)
Pregnancy	all ages	—	27
Breastfeeding	18 years and younger	—	10
Breastfeeding	19 years and older	—	9

vegetarians have higher RDAs in some cases to allow for lower absorption. These vegetarian RDAs can be reduced if vitamin C is included in the diet. It is not unusual for vegetarians to consume large amounts of vitamin C, which increases their absorption of dietary iron. Most grain products in the United States are fortified with iron. Despite this fortification, most pregnant and premenopausal women receive less than the RDA for iron in their diet.

Balancing Deficiency and Overload of Iron

Iron is an unusual nutrient because the body has a very limited capability to excrete excess iron. The three main factors that affect iron balance are absorption, losses, and the amount in stores. Iron is lost in very small amounts, mostly in blood and cells that are excreted into the intestinal tract.

The amount of iron in the body is largely controlled by varying the absorption rate. This varying absorption rate complicates the calculation of how much iron will be absorbed from a meal. In addition, there are enhancers and inhibitors of iron absorption in many meals. The RDAs for iron have been established to supply enough iron when there is a need for extra iron.

Several special proteins assist the body in absorbing iron from food. In the lining of the intestines is a protein called *mucosal ferritin*. Mucosal ferritin stores iron in the mucous cells lining the intestines, as seen in Figure 11-4. When the body needs iron, another protein, *mucosal transferrin*, receives iron bound to the mucosal ferritin and delivers it to the blood. The iron is then received by yet another protein, *transferrin*, which transports iron to the body in blood plasma. If iron is not needed, the iron in the mucous cells in the intestines is excreted after a few days when the cells are shed. In this way, dietary iron is held for use and released if not needed.

Iron is very efficiently recycled in the body. Red blood cells contain most of the iron in the body. After about four months, the older red blood cells die and the iron is recycled.

Iron is transported by transferrin in the plasma. *Transferrin receptors* are synthesized on cell surfaces when a cell needs iron. Transferrin receptors are also used to prevent free iron from appearing in the blood, which could cause free radical damage.

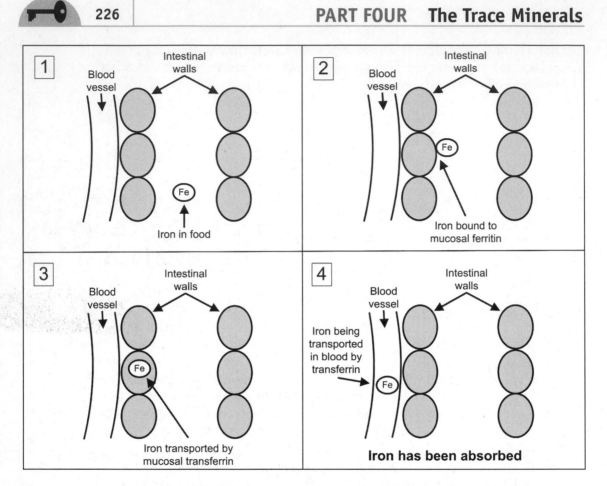

Figure 11-4   Iron absorption from the intestines into the blood.

Three Forms of Iron

Ferritin is used to store extra iron.
Transferrin transports iron in the blood plasma.
Hemosiderin is a compact storage of iron.

Iron is stored in the form of a protein called *ferritin*. Ferritin can store and release iron to meet demands. *Hemosiderin* can also store iron in the body. Hemosiderin is a protein larger than ferritin and it has a higher iron content. The

primary storage sites in the body for ferritin and hemosiderin are the liver, the spleen, and the bone marrow. This stored iron is especially important in the last six weeks of pregnancy.

Enhancers and Inhibitors of Iron Absorption

The iron absorbed from a meal can vary ten-fold depending on the need for iron and the enhancers and inhibitors of iron present in the meal. Adequate stomach acid is important for iron absorption. Iron is found in food in two types: *Heme iron* and *non-heme iron*.

HEME IRON

Heme iron is found only in meat, poultry, and fish. Heme iron is not found in other food; even dairy products and eggs have no heme iron. Heme iron accounts for less than 10 to 15 percent of the iron consumed, even in diets with high amounts of meat. Because heme iron is better absorbed, heme iron can account for up to 30 percent of the iron absorbed from a diet.

NON-HEME IRON

The other type of iron in food is called non-heme iron. Non-heme iron makes up about 90 percent of the iron in most diets. Non-heme iron is widely distributed in many foods; please refer to Graph 11-1. Good sources include green leafy vegetables, dried fruit, blackstrap molasses, nuts and seeds, and fortified grains. Cooking acidic food in cast iron pans can add iron to the diet. Iron supplements and iron used for food fortification are in the non-heme form. In addition, part of the iron in meat, poultry, and fish is non-heme iron.

Although it occurs in smaller amounts in the diet, heme iron is normally more easily absorbed than non-heme iron. While the absorption of non-heme iron is influenced by the other components of a meal, heme iron is much less influenced by accompanying enhancers or inhibitors. More iron is absorbed if iron stores in the body are low. Less iron is absorbed if iron stores in the body are high. The effect of body stores of iron on absorption is more pronounced with non-heme iron

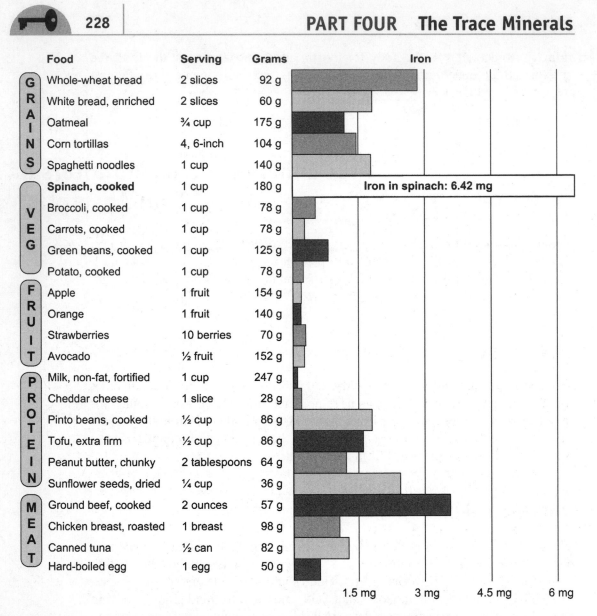

Food	Serving	Grams	Iron
GRAINS			
Whole-wheat bread	2 slices	92 g	
White bread, enriched	2 slices	60 g	
Oatmeal	¾ cup	175 g	
Corn tortillas	4, 6-inch	104 g	
Spaghetti noodles	1 cup	140 g	
VEG			
Spinach, cooked	1 cup	180 g	Iron in spinach: 6.42 mg
Broccoli, cooked	1 cup	78 g	
Carrots, cooked	1 cup	78 g	
Green beans, cooked	1 cup	125 g	
Potato, cooked	1 cup	78 g	
FRUIT			
Apple	1 fruit	154 g	
Orange	1 fruit	140 g	
Strawberries	10 berries	70 g	
Avocado	½ fruit	152 g	
PROTEIN			
Milk, non-fat, fortified	1 cup	247 g	
Cheddar cheese	1 slice	28 g	
Pinto beans, cooked	½ cup	86 g	
Tofu, extra firm	½ cup	86 g	
Peanut butter, chunky	2 tablespoons	64 g	
Sunflower seeds, dried	¼ cup	36 g	
MEAT			
Ground beef, cooked	2 ounces	57 g	
Chicken breast, roasted	1 breast	98 g	
Canned tuna	½ can	82 g	
Hard-boiled egg	1 egg	50 g	

1.5 mg   3 mg   4.5 mg   6 mg

Graph 11-1   Iron content of some common foods.

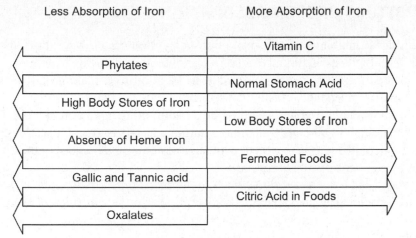

Figure 11-5   Enhancers and inhibitors of iron absorption.

than it is with heme iron. Several other factors can influence absorption of iron, as seen in Figure 11-5.

VITAMIN C ENHANCES IRON ABSORPTION

Vitamin C from food or from supplements is the strongest enhancer of iron absorption. The presence of vitamin C in a meal can increase the amount of non-heme iron that is absorbed. A meal with 25 mg of vitamin C can double the iron absorbed. A meal with 50 mg of vitamin C can increase non-heme iron absorption up to six times as much as a meal without vitamin C. Vitamin C changes the ferric iron (Fe^{3+}) in food into ferrous iron (Fe^{2+}). The vitamin C in the food becomes chelated with the iron in an *iron ascorbate* form that is very absorbable. However, vitamin C does not help with the absorption of iron from iron supplements.

THREE OTHER ENHANCERS OF IRON ABSORPTION

There are three other enhancers of iron absorption. The mechanisms of action and the amount of the increases in absorption of these enhancers are not known at present. First, the presence of heme iron acts as an enhancer of non-heme iron absorption. Second, fermented foods such as sauerkraut and fermented soy sauce can enhance the absorption of iron from a meal. Third, several food acids, such as citric acid, increase iron absorption.

INHIBITORS OF IRON ABSORPTION

The strongest inhibitors of non-heme iron absorption are *phytates*, also known as *phytic acid*. Phytates are found in legumes and whole grains. Small amounts of phytates can reduce iron absorption by half. In addition to containing phytates, soybeans have an independent factor that reduces the bioavailability of iron. Legumes, however, are rich in iron and can provide a useful amount of iron in a diet despite lower bioavailability. Foods containing vitamin C, such as green vegetables, are an especially useful addition to a meal with legumes or grains because of the ability of vitamin C to increase iron absorption.

Certain foods, such as raw spinach, contain *oxalates* that can bind some of their iron, making it unavailable for absorption. Certain *polyphenols* inhibit the bioavailability of iron. Many polyphenols such as *flavonoids* do not inhibit iron absorption. Only the polyphenols with *gallic acid*, which are often found in tannins, have been found to interfere with iron absorption. Iron absorption can be reduced by drinking beverages with tannic acid, such as wine, tea, or coffee during or within two hours of a meal. Spices such as oregano can also reduce iron absorption. Vitamin C greatly reduces the inhibiting effects of oxalates and tannins. Since many of the vegetables with oxalates also contain large amounts of vitamin C, this may compensate for the iron-inhibiting effects of the oxalates.

Iron Overload

Certain people, about one in 200, have a predisposition to iron overload called *hemochromatosis*. Iron overload is very rare in individuals without a genetic predisposition, even with prolonged iron supplementation. The carefully controlled absorption of iron prevents excess in most people. People may have hemochromatosis and not know it. For this reason, adult men and postmenopausal women are normally advised not to take iron supplements.

Children are at risk of accidental overdose from products containing iron. The largest cause of death from poisoning in children under six years of age is accidental iron overdose. Iron supplements, especially potent iron supplements meant for pregnant women, should be kept out of the reach of children.

Potent iron supplements may cause stomach and intestinal irritation and other digestive discomforts. Iron supplementation may cause constipation and cause the stools to appear darker. Liquid iron supplements can even stain teeth. Iron supplements should be taken with food to prevent discomfort.

Summary for Iron

Main functions: transportation of oxygen and energy
    metabolism.

RDA: adults, 7 to 18 mg; pregnant women, 27 mg.

Toxicity: only affects those with a genetic disorder.

Digestive discomfort may occur if taken on an empty
    stomach.

Tolerable upper intake level is set at 45 mg, 40 mg
    for ages under 14.

Deficiency can cause fatigue and anemia. Deficiency
    is common.

Sources: meat, fish, poultry, green leafy vegetables,
    dried fruit, blackstrap molasses, nuts and seeds,
    and fortified grains.

Forms in the body: ferritin, transferrin, hemosiderin.

To prevent gastrointestinal discomfort, the Food and Nutrition Board of the Institute of Medicine has established tolerable upper intake levels (UL). For children under 14 years of age, the UL is 40 mg of iron. For all others the UL is 45 mg of iron. For people with genetic predispositions to iron overload and those with alcoholic cirrhosis, the safe dose may be lower. Higher doses may be prescribed for pregnant women.

There has been some association between the intake of heme iron and the risk of heart attacks. When iron stores are high, heme iron absorption is not as effectively limited as is the absorption of non-heme iron. This may lead to a higher risk of heart attacks because of the free radical action of excess iron on LDL. The possibility of excess heme iron from animal sources increasing the risk of heart attacks is being investigated.

Iron Supplements

Absorption of iron from supplements depends on the form of the iron. The forms that are best absorbed and cause the least gastrointestinal discomfort are ferrous ascorbate and ferrous gluconate. Ferrous sulfate monohydrate is also well absorbed. The forms of iron supplements that are least well absorbed include ferrous fumarate, ferrous succinate, and ferric saccharate.

To summarize, iron is needed to transport and store oxygen in the body. Iron is also essential in energy production in the cell. Iron deficiency is the most common nutrient deficiency in the world.

Quiz

Refer to the text in this chapter if necessary. A good score is at least 8 correct answers out of these 10 questions. The answers are listed in the back of this book.

1. Iron is needed for:
   (a) Transport of oxygen in blood.
   (b) Storage of oxygen in muscles.
   (c) Energy production in the electron transport chain.
   (d) All of the above.

2. Which is NOT part of the electron transport chain?
   (a) Succinate dehydrogenase.
   (b) Pyruvic acid.
   (c) Iron-sulfur proteins.
   (d) Cytochromes.

3. Iron catalyzes reactions that turn hydrogen peroxide into:
   (a) Carbon.
   (b) Nitrogen.
   (c) Water.
   (d) Iron.

4. In response to an infection, iron in blood:
   (a) Is decreased.
   (b) Stays the same.
   (c) Is increased.
   (d) Is greatly increased.

5. Which can interfere with iron absorption?
   (a) Calcium.
   (b) Zinc.
   (c) Vitamin A.
   (d) Copper.

6. Which group is NOT at risk for iron deficiency anemia?
   (a) Pregnant women.
   (b) Children between 6 months and 4 years of age.
   (c) Adolescent girls.
   (d) Adults over 50 years of age.

7. Which form of iron is used in transport?
   (a) Ferritin.
   (b) Hemosiderin.
   (c) Transferrin.
   (d) None of the above.

8. Lowered absorption of iron is caused by:
   (a) Vitamin C.
   (b) Phytates.
   (c) Low iron stores in the body.
   (d) Fermented foods.

9. Iron absorption is NOT reduced by:
   (a) Wine.
   (b) Coffee.
   (c) Lettuce.
   (d) Tea.

10. Which form of iron found in supplements is poorly absorbed?
    (a) Ferrous fumarate.
    (b) Ferrous ascorbate.
    (c) Ferrous gluconate.
    (d) Ferrous sulfate monohydrate.

CHAPTER 12

Zinc

The Growth Mineral

Zinc is a trace mineral that is essential for all forms of life, including plants, animals, and microorganisms. The chemical symbol for zinc is Zn. Clinical zinc deficiency was first reported in 1961, when it was discovered that certain children absorbed so little zinc that they failed to grow properly. Zinc plays important roles in growth and development, neurological function, the immune system, and in reproduction. Over 85 percent of the total body zinc is found in skeletal muscle and bone. Zinc found in blood plasma comprises only one-tenth of one percent of total body zinc.

Zinc and Enzymes

Zinc is needed for the activity of nearly 100 enzymes. Enzymes that use a metal ion such as zinc as a cofactor are called *metalloenzymes*. Please refer to Figure 12-1 to see how these metalloenzymes fit together. Zinc's action as an electron acceptor

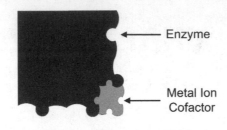

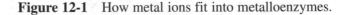

Metal ion cofactors include:
cobalt, copper, iron, manganese,
molybdenum, nickel, and zinc.

Figure 12-1   How metal ions fit into metalloenzymes.

contributes to the catalytic activity of many of these enzymes. In another role, zinc is important for the synthesis, storage, and release of insulin in the pancreas.

Zinc Finger-Like Structures

Zinc stabilizes the structure of a number of proteins. Zinc helps certain proteins to fold by attaching to the amino acids *cysteine* and *histidine*, as seen in Figure 12-2.

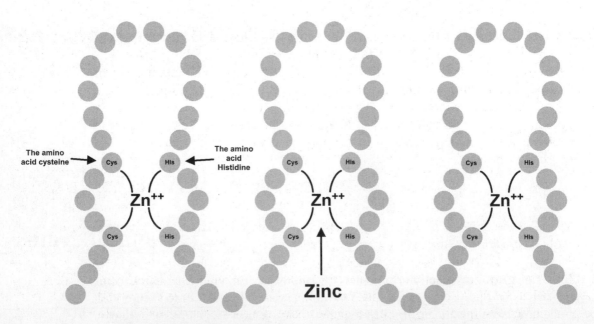

Figure 12-2   Zinc can stabilize the structure of some proteins.

These folded proteins have a *zinc finger-like structure* that increases their structural stability. Proteins that use zinc finger-like structures include retinal (vitamin A) receptors in the eye and vitamin D receptors. Zinc finger-like structures are important for proteins that regulate hormones, such as testosterone. Zinc is also found in the proteins of viruses; one example is the human immunodeficiency virus.

Zinc's Antioxidant Role

Zinc is an important component of *superoxide dismutase* (SOD). The free radical *Superoxide* is made of two paired oxygen atoms with an extra electron. Superoxide dismutase is an important antioxidant enzyme that converts superoxide free radicals into oxygen and hydrogen peroxide. Hydrogen peroxide is less dangerous as a free radical and can be further degraded into water and oxygen.

There are three types of superoxide dismutase. Two types of superoxide dismutase use zinc for structural stability and copper for catalytic activity. The third type of superoxide dismutase is found in the mitochondria and has manganese as its center. In addition, cell membranes need zinc for antioxidant protection and for structural integrity.

Zinc Assists Cell Signaling

Zinc finger proteins can bind to DNA to influence which genes are expressed. Zinc influences hormone release and nerve impulse transmission by assisting cell signaling.

Nutrient Interactions with Zinc

Zinc can reduce the absorption of copper. Typical intakes of zinc do not interfere with copper bioavailability. However, large quantities of supplemental zinc, well over the tolerable upper intake level (UL), can reduce the bioavailability of copper in the body. Too much zinc can cause the synthesis of a copper-binding protein called *metallothionein* in the intestines. This metallothionein traps copper and prevents its absorption. High copper intakes have not, however, been found to affect zinc absorption.

Zinc absorption may be influenced by iron. Higher levels of supplementary iron may decrease zinc absorption. Iron in food has not been shown to interfere with zinc absorption. Pregnant and breastfeeding women who take more than 60 mg of elemental iron per day may need supplemental zinc because of this interaction.

Calcium may reduce zinc absorption in diets high in phytates. Plants store phosphates in *phytates (phytic acid)*. Tortillas are high in phytates and are made with calcium oxide (lime). In diets consisting largely of tortillas, a slight increase in zinc consumption may be recommended to offset the slight decrease in zinc bioavailability.

Zinc is needed to produce an active form of vitamin A, retinal, which is used in visual pigments. Zinc is also needed by the retinol-binding protein that transports vitamin A. Zinc deficiency can cause signs of vitamin A deficiency, even when there is sufficient vitamin A.

Severe Zinc Deficiency

Severe zinc deficiency is normally seen only in individuals with genetic disorders. Dietary deficiency of zinc is unlikely to cause severe zinc deficiency. However, prolonged diarrhea or severe burns can cause severe zinc deficiency.

Mild Zinc Deficiency

Mild zinc deficiency is commonly found in children in developing countries. Mild zinc deficiency can impair weight gain and can prevent children from growing taller. Researchers suspect that lowered zinc levels can interfere with the cellular response to the growth-regulating hormone *insulin-like growth factor-1 (IGF-1)*. Zinc supplementation has been found to correct these growth problems.

INFANT ZINC DEFICIENCY

Infants fed cow's milk may be more susceptible to mild zinc deficiency. Thirty-two percent of the zinc in cow's milk is bound to casein and the majority of the remaining zinc (63 percent) is bound to colloidal calcium phosphate, reducing bioavailability. Zinc bioavailability is even lower in soy-based formulas. Breast milk has excellent bioavailability of zinc, although zinc content falls off after the first six months of breastfeeding.

CHILD ZINC DEFICIENCY

Infectious diarrhea results in the deaths of millions of children each year. Mild zinc deficiency can increase the susceptibility of children to infectious diarrhea. Unfortunately, diarrhea reduces zinc absorption, worsening the problem. Zinc supplementation can be added to oral rehydration therapy to significantly increase survival with persistent childhood diarrhea. Adequate zinc can also reduce the effect of bacterial toxins on the intestines.

Younger zinc-deficient children may also be more susceptible to infections. The incidence of pneumonia in children in developing countries has been reduced with zinc supplementation. Zinc is essential for T-lymphocyte development and activation.

ZINC AND PREGNANCY

Four out of five pregnant women worldwide have inadequate zinc levels. Low maternal zinc status has been associated with low birth weight and premature delivery. Mothers with low zinc levels also have more labor and delivery complications than mothers with normal zinc levels. Zinc supplementation has been found to help pregnant women who have low nutritional status. Zinc supplementation has not been as effective with pregnant women in the developed world where nutrient intakes are generally more adequate.

ZINC AND AGE-RELATED MACULAR DEGENERATION

Zinc supplementation has been found to help reduce the risk of age-related macular degeneration in some studies. Other studies have not shown a connection between dietary zinc intake and the incidence of age-related macular degeneration. Zinc is found in high concentrations in the macula (central) portion of the retina. Zinc concentration in this area of the eye has been found to decrease with aging. Current research is trying to find out if zinc and antioxidant supplementation can help reduce the effects of age-related macular degeneration.

ZINC AND HIV

Individuals with *human immunodeficiency virus* (HIV) are particularly susceptible to zinc deficiency. Lowered blood levels of zinc have been correlated with increased death rates in HIV patients. Zinc supplementation can reduce opportunistic infection in those with HIV. However, the HIV virus also needs zinc. Zinc supplements may in fact backfire to decrease survival time and stimulate the virus. Helping the immune system get adequate zinc without helping the HIV will require further research.

ZINC AND OLDER ADULTS

Older adults have average zinc intakes that tend to be lower than the RDA. To avoid impaired immune system functioning, older adults should be sure to maintain adequate zinc intake.

ZINC AND ALCOHOLISM

Alcoholics are at increased risk of zinc deficiency from both impaired absorption and increased urinary losses. One-third to one-half of alcoholics have been found to have low zinc levels.

Individuals at Greater Risk of Zinc Deficiency

Children and infants
Pregnant and breastfeeding women, especially teens
People whose main diet consists of grains and legumes
People with chronic digestive disorders
Patients receiving only intravenous feedings
Alcoholics
People with sickle cell anemia
Adults 65 years and older
Malnourished people or those with anorexia nervosa

Food Sources of Zinc

Oysters are the richest source of zinc; in fact, six oysters have more than the tolerable upper intake level (UL) of zinc. Animal products have the highest levels of zinc and also have a high bioavailability; please refer to Graph 12-1. Beans are a good source of zinc, but bioavailability may be lower because the phytates in

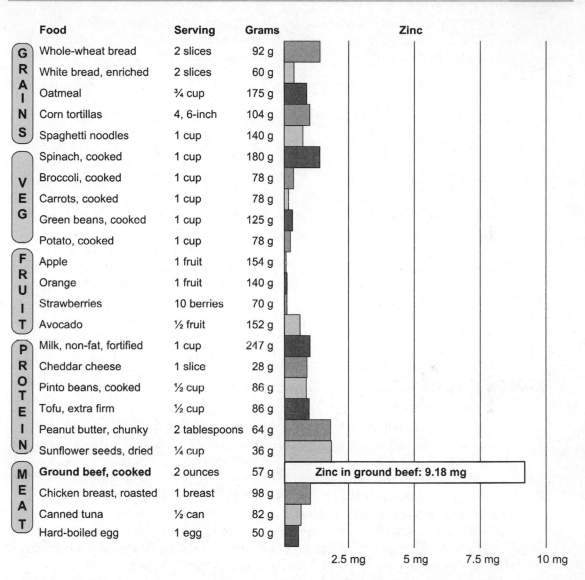

Food	Serving	Grams	Zinc
GRAINS			
Whole-wheat bread	2 slices	92 g	
White bread, enriched	2 slices	60 g	
Oatmeal	¾ cup	175 g	
Corn tortillas	4, 6-inch	104 g	
Spaghetti noodles	1 cup	140 g	
VEG			
Spinach, cooked	1 cup	180 g	
Broccoli, cooked	1 cup	78 g	
Carrots, cooked	1 cup	78 g	
Green beans, cooked	1 cup	125 g	
Potato, cooked	1 cup	78 g	
FRUIT			
Apple	1 fruit	154 g	
Orange	1 fruit	140 g	
Strawberries	10 berries	70 g	
Avocado	½ fruit	152 g	
PROTEIN			
Milk, non-fat, fortified	1 cup	247 g	
Cheddar cheese	1 slice	28 g	
Pinto beans, cooked	½ cup	86 g	
Tofu, extra firm	½ cup	86 g	
Peanut butter, chunky	2 tablespoons	64 g	
Sunflower seeds, dried	¼ cup	36 g	
MEAT			
Ground beef, cooked	2 ounces	57 g	Zinc in ground beef: 9.18 mg
Chicken breast, roasted	1 breast	98 g	
Canned tuna	½ can	82 g	
Hard-boiled egg	1 egg	50 g	

2.5 mg   5 mg   7.5 mg   10 mg

Graph 12-1   Zinc content of some common foods.

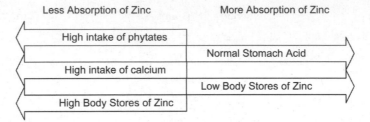

Figure 12-3 Some factors affecting zinc absorption.

beans bind some of the zinc, as seen in Figure 12-3. Dairy products and grains are moderate sources of zinc.

Leavened whole grains have more available zinc because the enzymatic action of yeast breaks down phytates. Enriched white flour has less than a quarter of the zinc found in whole wheat flour because the zinc is lost in milling. Many breakfast cereals are fortified with zinc. Zinc deficiency in agricultural samples is very common with only a small percentage of agricultural soils providing adequate zinc.

Absorption of Zinc

Zinc absorption in the intestines varies from 15 to 40 percent. Zinc can be absorbed by a special protein in the intestinal cells called metallothionein. As you may recall, this is the same protein that can bind copper. Zinc is held by metallothionein until needed in the blood. If the zinc is not needed, it may be eliminated in the stool when the intestinal cells slough off. Metallothionein also binds zinc in the liver until the zinc is needed. Some zinc is transported by the iron transporter *transferrin*. Zinc enters the blood surrounding the digestive system and is commonly bound to *albumin*, a blood protein. Iron and zinc should be somewhat balanced in the diet so that one does not interfere with the other's absorption.

Considerable amounts of zinc enter the intestines in pancreatic juices and intestinal cell secretions, as seen in Figure 12-4. The intestines are given the opportunity to reabsorb this recycled zinc along with dietary zinc, or to allow the zinc to be eliminated in the stool. Both zinc recycled from the body and zinc from food may bind to dietary amino acids, *peptides*, nucleic acids, and phytates in the intestines. Zinc also binds tenaciously to the protein *casein* in dairy products. Most of this bound zinc passes in the stool. Urine losses are usually under 10 percent of the total zinc eliminated. Small amounts of zinc are also lost from the body in skin cells, the menstrual cycle, sweat, semen, and hair.

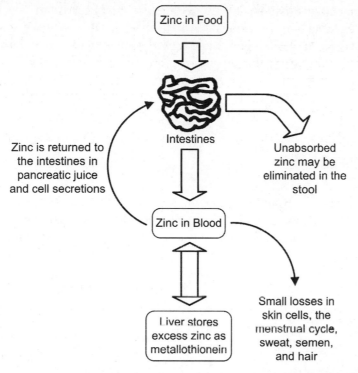

Zinc in Food

Intestines

Zinc is returned to
the intestines in
pancreatic juice
and cell secretions

Unabsorbed
zinc may be
eliminated in the
stool

Zinc in Blood

Small losses in
skin cells, the
menstrual cycle,
sweat, semen,
and hair

Liver stores
excess zinc as
metallothionein

Figure 12-4   Zinc is returned to the intestines.

Summary for Zinc

Main functions: in enzymes, with hormones, protein structure, and diverse functions.

RDA: 2 to 6 mg for children and 8 to 13 mg for adolescents and adults.

Toxicity: excesses may induce a copper deficiency and gastrointestinal disturbances.

Tolerable upper intake level is 3 mg for infants ranging up to 40 mg for adults.

Deficiency can cause growth retardation and susceptibility to infection in children.

Sources: meat, fish, poultry, whole grains, and Brazil nuts.

Forms in the body: found in zinc-copper superoxide dismutase. Can be bound to metallothionein or albumin.

Certain vegetarians may need increased dietary zinc because of lower absorption. Most vegetarians have adequate zinc to meet needs. Vegetarians or others whose major food staples are grains and legumes may consume enough phytates to reduce absorption by one-third. The consumption of dairy products also reduces zinc bioavailability because of the zinc-binding action of casein and calcium.

Zinc Supplements

Zinc supplements are available in a number of forms. Zinc gluconate is readily absorbable. Zinc picolinate is also absorbable, but absorption may be offset by increased elimination. Zinc sulfate and zinc acetate are also available as supplements.

Zinc Requirements

The RDAs have been set for zinc at levels believed to prevent deficiency in most people, as seen in Table 12-1. Adequate intakes (AI) have been set for younger

Table 12-1   RDAs and adequate intakes (AI) for zinc for all ages.

RDAs for Zinc	Age	Males mg/day	Females mg/day
Infants	0–6 months	2 (AI)	2 (AI)
Infants	7–12 months	3	3
Children	1–3 years	3	3
Children	4–8 years	5	5
Children	9–13 years	8	8
Adolescents	14–18 years	11	9
Adults	19 years and older	11	8
Pregnancy	18 years and younger	—	12
Pregnancy	19 years and older	—	11
Breastfeeding	18 years and younger	—	13
Breastfeeding	19 years and older	—	12

infants. The optimal amount of supplemental zinc for disease resistance for adults may be 35 mg, just under the upper intake level of 40 mg.

Excess Zinc

The major problem with excess zinc consumption is copper deficiency. A tolerable upper intake level (UL) has been established to prevent a reduction in the antioxidant (*copper-zinc superoxide dismutase*) activity in red blood cells. This UL includes zinc from dietary, fortified, and supplementary sources added together. The UL for infants is 4 to 5 mg. For children one to three years of age, the UL is 7 mg. For children four to eight years of age, the UL is 12 mg. For children nine to thirteen years of age, the UL is 23 mg. For adolescents, the UL is 34 mg. For adults, the UL is 40 mg.

ZINC LOZENGES

Many people use zinc lozenges to treat cold symptoms. There are still questions about the effectiveness of these lozenges. Zinc intakes well above the tolerable upper intake level (UL) can result from frequent use of zinc lozenges. Short-term use of zinc lozenges has not been found to cause problems from excesses of zinc. However, frequent use of zinc lozenges for over six weeks can result in copper deficiency. Use of zinc sprays in the nose has the potential to cause loss of smell. Since this type of loss of smell is permanent, nasal zinc sprays and gels should be avoided.

ZINC CONTAMINATION

Supplemental zinc above the UL of 40 mg has caused mild gastrointestinal distress. Acute zinc toxicity can cause diarrhea, abdominal pain, nausea, and vomiting. Zinc toxicity has occurred from food and beverages contaminated from storage in galvanized containers. Zinc fumes, such as those inhaled by welders of galvanized metal, have also caused zinc toxicity.

To summarize, zinc is a versatile nutrient with many functions. Zinc is needed by metalloenzymes such as the enzymes that synthesize insulin. Zinc is used to fold and shape certain proteins.

Quiz

Refer to the text in this chapter if necessary. A good score is at least 8 correct answers out of these 10 questions. The answers are listed in the back of this book.

1. Most of the zinc in the body is found in:
    (a) The pancreas.
    (b) The liver.
    (c) Skeletal muscle and bone.
    (d) The brain.

2. Proteins with zinc finger-like structures are important for all of the following EXCEPT:
    (a) Energy production.
    (b) Regulation of the hormone testosterone.
    (c) Retinoic acid activity in the eye.
    (d) Vitamin D activity.

3. Very high intakes of zinc can interfere with:
    (a) Magnesium absorption.
    (b) Copper absorption.
    (c) Vitamin A absorption.
    (d) Vitamin E absorption.

4. Signs of childhood zinc deficiency are:
    (a) Increased susceptibility to infectious diarrhea.
    (b) Increased susceptibility to infections.
    (c) Failure to grow.
    (d) All of the above.

5. The lowest content of dietary zinc is found in:
    (a) Fruit.
    (b) Legumes.
    (c) Dairy products.
    (d) Meats.

6. Phytates in grains and beans:
   (a) Increase the bioavailability of zinc.
   (b) Decrease the bioavailability of zinc.
   (c) Do not affect the bioavailability of zinc.
   (d) None of the above.

7. Zinc can enter the intestines from all of the following EXCEPT:
   (a) Pancreatic juices.
   (b) Intestinal cell secretions.
   (c) Gall bile.
   (d) Food intake.

8. Metallothionein:
   (a) Can bind zinc for later absorption.
   (b) Can bind copper.
   (c) Is found in intestinal cells.
   (d) All of the above.

9. The best form of supplemental zinc is:
   (a) Zinc gluconate.
   (b) Zinc acetate.
   (c) Zinc picolinate.
   (d) Zinc sulfate.

10. The upper level of intake (UL) for adults for zinc is:
    (a) 5 mg.
    (b) 20 mg.
    (c) 40 mg.
    (d) 60 mg.

CHAPTER 13

Minor Trace Minerals

Iodine, Selenium, Copper, Manganese, Fluoride, Chromium, Molybdenum, Lead, and Mercury

Iodine

Iodine is needed to make thyroid hormones. Iodine is a non-metallic trace element that is essential for nutrition. The chemical symbol for iodine is the letter I. *Iodine* is the name of the element in food and *iodide* is the name of the ion form found in the body. Only tiny amounts of dietary iodine are needed.

Over geologic time, iodine has washed out of soils and into the ocean. Seaweed and fish acquire iodine from the ocean; these are the richest dietary sources of iodine. The soils of mountainous areas and flooded plains are often deficient in iodine.

THYROID HORMONES AND ENERGY PRODUCTION

Thyroid hormones regulate the burning of energy in the body by controlling the rate that oxygen burns in the cells. This regulates the metabolism of fats, proteins, and carbohydrates. Thyroid hormones increase the burning of fats and regulate the burning of carbohydrates. Overall, thyroid hormones increase the production of energy in the body.

IODIDE IN THYROID HORMONES

There are two important thyroid hormones that regulate growth, metabolism, and reproduction. The thyroid gland takes iodide from the blood and converts it into *thyroxine*. Thyroxine is known as *T4* because it has four iodide ions. Thyroxine is stored in the thyroid gland and released into the blood as needed. Thyroxine can be converted into the active form of thyroid hormone called *triiodothyronine*. Triiodothyronine is also known as *T3* because it has three iodide ions. This conversion of thyroxine to the active form, T3, takes place in tissues such as the liver and the brain and requires selenium. The action of T3 is based upon its capability of causing the synthesis of specific proteins in the nucleus of the cell.

> ### Thyroid Hormones with Iodide
>
> *Thyroxine* The storage and transport thyroid hormone known as T4
> *Triiodothyronine* The active thyroid hormone known as T3

There are several steps in the regulation of the thyroid hormones. Deep in the brain is the *hypothalamus*. When thyroid hormones are needed, the hypothalamus secretes a hormone (*thyrotropin-releasing hormone*) that targets the *pituitary gland*. In response, the pituitary gland secretes *thyroid-stimulating hormone* (*TSH*), as seen in Figure 13-1. This thyroid-stimulating hormone tells the thyroid gland to take more iodide from the blood. The thyroid-stimulating hormone also stimulates the thyroid gland to produce and release more thyroxine and small amounts of the active thyroid hormone T3.

When levels of thyroxine are adequate, the pituitary gland secretes less of the thyroid-stimulating hormone. When levels of thyroxine start falling, the pituitary gland secretes more of the thyroid-stimulating hormone.

Iodine deficiency causes lowered levels of thyroxine in the blood. In response, the pituitary gland releases more thyroid-stimulating hormone. Over time, this can

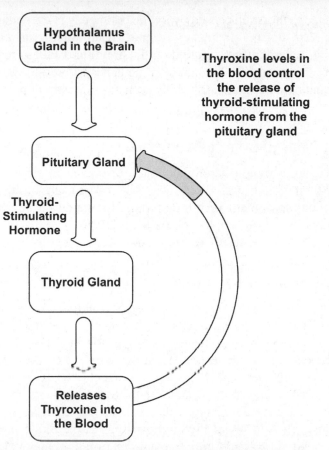

Figure 13-1   Control of thyroid hormones.

lead to enlargement of the thyroid gland, located in the neck. This enlargement of the thyroid gland is called *goiter* and can result in a swelling in the thyroid gland in the neck.

IODINE DEFICIENCY

Iodine deficiency causes problems at all stages of life. Iodine is needed in the diet so that the body can produce thyroid hormones. Thyroid hormones help develop the nervous system by aiding the formation of the myelin sheath of certain nerves in the central nervous system. These nerve sheaths form before and shortly after birth. Iodine deficiency can cause brain damage, especially in the children of pregnant women after the first trimester and in children of up to three years of age. Iodine is critical for the growth and development of the brain and central nervous system.

The damage to the brain caused by iodine deficiency is irreversible. If the deficiency is severe during pregnancy, it may result in *cretinism* in the child. Cretinism is a state of stunted growth and extreme mental retardation, resulting in intelligence quotients as low as 20. Pregnant and breastfeeding women can ensure adequate iodine by taking a daily prenatal supplement providing 150 mcg of iodine.

Brain Damage from Iodine Deficiency

Iodine deficiency is the most common cause of preventable brain damage in the world. Diets that do not include seaweed, fish, iodized salt, or other iodized food have been found to contain very little iodine. Iodine deficiency disorders affect three-quarters of a billion people worldwide. It is estimated that 50 million people have some brain damage resulting from iodine deficiency.

Goiter and Hypothyroidism

One of the earliest signs of iodine deficiency is goiter, an enlargement of the thyroid gland. In older children and adults, goiter may be reversed with adequate intake of iodine. The incidence of goiter is more common in adolescent girls. Children who are deficient in iodine have poorer school performance, more learning disabilities, and lower intelligence quotients than normal children. Childhood iodine deficiency can cause an average lowering of intelligence quotients by 13 points.

More severe iodine deficiency can result in *hypothyroidism*. Symptoms of hypothyroidism (low levels of thyroid hormones in the blood) include dry skin, swellings around the lips and nose, mental deterioration, and a slow basal metabolic rate.

Summary for Iodine

Main functions: thyroid functions.

RDA: adults, 150 mcg.

Toxicity: low toxicity, excesses may cause a rise in thyroid stimulating hormone.

Tolerable upper intake level is 1100 mcg for adults.

Deficiency can cause goiter, cretinism, and brain damage to fetuses.

Sources: iodized salt, seaweed, fish.

Forms in the body: thyroxine (T4) and the more active form, triiodothyronine (T3).

Causes of Iodine Deficiency

One cause of iodine deficiency is the over-consumption of certain plants that reduce thyroid hormones even if dietary iodine is adequate. Over-consumption of plants in the cabbage family and certain other plants can also cause goiter. These plants are known as *goitrogens* because they promote goiter.

Radiation and Iodine

Iodine deficiency can result in an increased susceptibility to thyroid cancer in populations exposed to radiation. Deficiency results in an increased uptake of iodide by the thyroid gland. The thyroid gland is also capable of storing ions of the radioactive form of iodide (iodine I-131).

IODINE IN FOOD

The amount of iodine needed at different ages is well understood. Iodized salt has eliminated most of the iodine deficiency in the United States and Canada where intakes are now generally found to be adequate. About 11 percent of Americans are low in iodide, and about seven percent of pregnant women are deficient. Seaweed and fish are valuable sources of trace minerals including iodine. Fast foods, breads, and dairy products all may have iodine added to them during production. Please refer to Table 13-1 for RDAs and adequate intakes (AI) for iodine.

Dietary iodine is converted into the iodide ion before it is absorbed. The iodide is totally absorbed. The iodide not needed by the thyroid gland is cleared by the kidneys.

Table 13-1   RDAs and adequate intakes (AI) for iodine for all ages.

RDAs for Iodine	Age	Males (mcg/day)	Females (mcg/day)
Infants	0–6 months	110 (AI)	110 (AI)
Infants	7–12 months	130 (AI)	130 (AI)
Children	1–3 years	90	90
Children	4–8 years	90	90
Children	9–13 years	120	120
Adolescents	14–18 years	150	150
Adults	19 years and older	150	150
Pregnancy	all ages	—	220
Breastfeeding	all ages	—	290

Iodine Supplements and Fortification

Iodine can be found in many multivitamin/multimineral supplements, usually in the form of potassium iodine. An extra 150 mcg of iodine each day has not been found to cause toxicity problems, even in addition to iodized salt and dietary iodine. Tolerable upper intake levels (UL) have been set for iodine to prevent high blood levels of thyroid stimulating hormone. The UL for children ages one to three is 200 mcg; for children ages four to eight, 300 mcg. For children ages nine to thirteen the UL is 600 mcg. For adolescents, the UL is 900 mcg and for adults it is 1100 mcg. The average intake for Japanese people is 2000-3000 mcg daily without causing any apparent problems. Higher levels of iodine are used to treat fibrocystic breast conditions under medical supervision.

To summarize, iodine, in its role in thyroid hormones, is vital for metabolism. Iodine is needed to prevent brain damage to fetuses.

Selenium

Selenium is a trace mineral that is essential in tiny amounts, but is toxic in larger amounts. The chemical symbol for selenium is *Se*. Some geographic areas have selenium in the soil and others are deficient. Plants do not appear to need selenium for survival. If selenium is present in soil, plants will take up the selenium.

Selenium is used in the body in antioxidant enzymes, by the immune system, and in conjunction with iodine to control metabolism. Approximately 30 percent of tissue selenium is found in the liver, 15 percent in kidneys, 30 percent in muscle, and 10 percent in blood plasma, as seen in Figure 13-2.

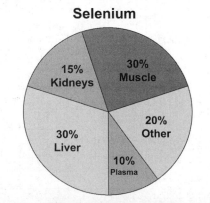

Selenium

15% Kidneys · 30% Muscle · 20% Other · 30% Liver · 10% Plasma

Figure 13-2 Where selenium is found in the body.

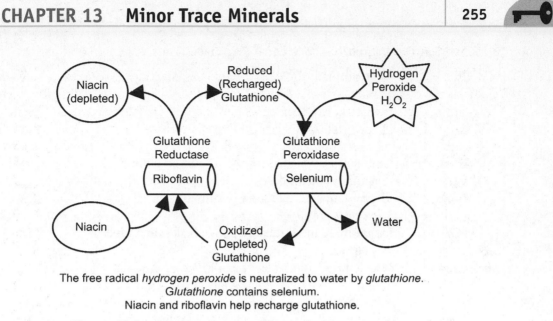

The free radical *hydrogen peroxide* is neutralized to water by *glutathione*.
Glutathione contains selenium.
Niacin and riboflavin help recharge glutathione.

Figure 13-3   Selenium is part of the antioxidant glutathione.

SELENIUM AS AN ANTIOXIDANT

Selenium is part of an important antioxidant, *glutathione peroxidase*. Glutathione peroxidase protects against oxidation inside cells, in cell membranes with vitamin E, in blood plasma, in sperm, and in the intestines. Glutathione peroxidase has the ability to transform harmful antioxidants such as hydrogen peroxide into water. Glutathione is recharged with the assistance of the B vitamins riboflavin and niacin, as seen in Figure 13-3.

Selenium is also found in another antioxidant, *selenoprotein P*. Selenoprotein P is capable of neutralizing nitrogen free radicals in the lining of blood vessels. Other minerals that are necessary components of antioxidant enzymes include copper, zinc, and iron.

SELENIUM AND THYROID HORMONES

Thyroid hormones exist as a storage form, thyroxine, with four iodide ions (T4), and as the active form with three iodide ions (T3). An enzyme that requires selenium removes the extra iodine ion to transform T4 into T3. Without selenium, the active form of thyroid hormone cannot be made. Thus, selenium and iodine are both necessary for proper thyroid function.

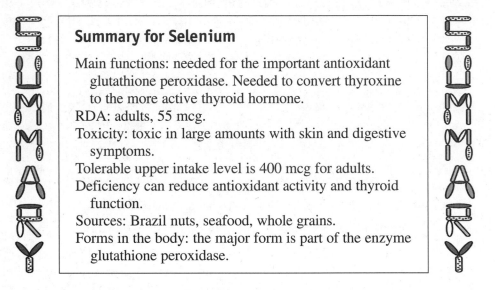

Summary for Selenium

Main functions: needed for the important antioxidant glutathione peroxidase. Needed to convert thyroxine to the more active thyroid hormone.
RDA: adults, 55 mcg.
Toxicity: toxic in large amounts with skin and digestive symptoms.
Tolerable upper intake level is 400 mcg for adults.
Deficiency can reduce antioxidant activity and thyroid function.
Sources: Brazil nuts, seafood, whole grains.
Forms in the body: the major form is part of the enzyme glutathione peroxidase.

SELENIUM AND THE IMMUNE SYSTEM

Selenium is needed by important immune system cells known as *T cells*. Selenium helps the T cells produce *cytokines*, which are used as messengers between and inside the cells. Adequate selenium may be needed to increase resistance to HIV infection. Selenium is also needed by white blood cells to fight microorganisms. Low dietary intakes of selenium may make people more vulnerable to *Keshan* disease. Keshan disease can cause heart problems.

SELENIUM AND CANCER

Excesses of selenium create *methylated selenium* in the body. This methylated form of selenium has been found to reduce tumor risk. Conversely, populations in areas with low selenium in the soil have been found to have higher cancer incidence. Supplementation with selenium has been found to decrease the risk of prostate cancer. Unfortunately, the risk of one type of skin cancer (squamous cell carcinoma) was found to be increased by selenium supplementation. Further research is needed to clarify selenium's role in cancer prevention.

SOURCES OF SELENIUM

Selenium is found in some soils in the United States and Canada. Since vegetables and grains are transported from different areas, people generally receive enough selenium in their diet. Brazil nuts are especially high in selenium, although they do vary in their content. Food sources of selenium can be seen in Graph 13-1. Infant formulas based upon cow's milk may be deficient in selenium.

Selenium supplements are available in several forms. *Sodium selenate* is very well absorbed, but much of it is lost in urine before it can be used. Only about half of the selenium in *sodium selenite* is absorbed. *Selenomethionine* occurs naturally in foods and is about 90 percent absorbed. Special selenium-fortified yeast can also contain selenomethionine. RDAs and adequate intakes (AI) for selenium are shown in Table 13-2.

TOXICITY OF EXCESS SELENIUM

Doses of over 1000 micrograms (mcg) daily of selenium can be toxic. Symptoms of toxicity can include brittleness and loss of hair and nails, vomiting and diarrhea, and skin problems. Tolerable upper intake levels (UL) have been set for selenium. For infants, the UL is 45 to 60 mcg. For children, the UL ranges from 90 mcg for ages one to three to 280 mcg for ages nine to thirteen. For adolescents and adults the UL is 400 mcg of selenium.

To summarize, selenium is important for antioxidant activity in the body. Selenium is also needed to transform thyroid hormone into its active form.

Table 13-2   RDAs and adequate intakes (AI) for selenium for all ages.

RDAs for Selenium	Age	Males (mcg/day)	Females (mcg/day)
Infants	0–6 months	15 (AI)	15 (AI)
Infants	7–12 months	20 (AI)	20 (AI)
Children	1–3 years	20	20
Children	4–8 years	30	30
Children	9–13 years	40	40
Adolescents	14–18 years	55	55
Adults	19 years and older	55	55
Pregnancy	all ages	—	60
Breastfeeding	all ages	—	70

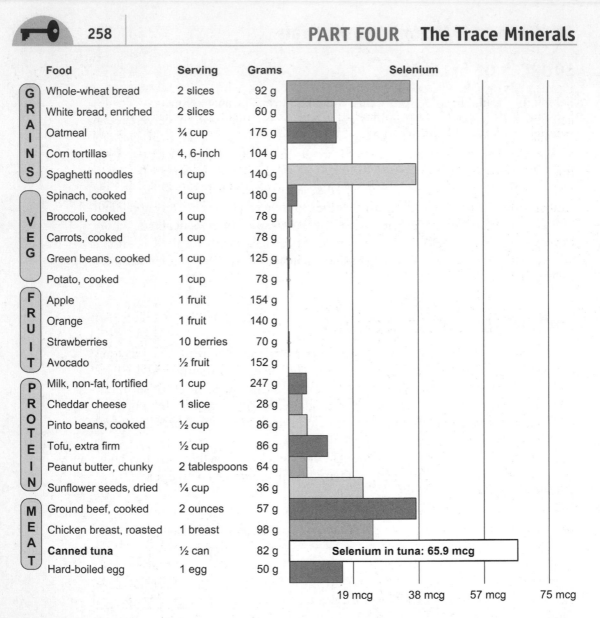

Food	Serving	Grams	Selenium
GRAINS			
Whole-wheat bread	2 slices	92 g	
White bread, enriched	2 slices	60 g	
Oatmeal	¾ cup	175 g	
Corn tortillas	4, 6-inch	104 g	
Spaghetti noodles	1 cup	140 g	
VEG			
Spinach, cooked	1 cup	180 g	
Broccoli, cooked	1 cup	78 g	
Carrots, cooked	1 cup	78 g	
Green beans, cooked	1 cup	125 g	
Potato, cooked	1 cup	78 g	
FRUIT			
Apple	1 fruit	154 g	
Orange	1 fruit	140 g	
Strawberries	10 berries	70 g	
Avocado	½ fruit	152 g	
PROTEIN			
Milk, non-fat, fortified	1 cup	247 g	
Cheddar cheese	1 slice	28 g	
Pinto beans, cooked	½ cup	86 g	
Tofu, extra firm	½ cup	86 g	
Peanut butter, chunky	2 tablespoons	64 g	
Sunflower seeds, dried	¼ cup	36 g	
MEAT			
Ground beef, cooked	2 ounces	57 g	
Chicken breast, roasted	1 breast	98 g	
Canned tuna	½ can	82 g	Selenium in tuna: 65.9 mcg
Hard-boiled egg	1 egg	50 g	

19 mcg   38 mcg   57 mcg   75 mcg

Graph 13-1   Selenium content of some common foods.

Copper

Copper is an essential trace mineral in animals and many plants. The chemical symbol for copper is *Cu* from the Latin name for copper, *cuprum*. There is less than one-tenth of one gram of copper in the human body. Copper can easily accept and donate electrons in the body. Copper can shift between the *cuprous* state with a single positive charge (Cu^+) to the *cupric* state, which has two positive charges (Cu^{++}). Copper has important roles in energy production in the cell and in scavenging free radicals.

COPPER IN ENERGY PRODUCTION

Copper is used in an enzyme called *cytochrome c oxidase* in the energy-producing mitochondria in the cell. Cytochrome c oxidase helps to make ATP, the energy battery of the cell. By catalyzing the reaction between molecular oxygen and water, cytochrome c oxidase generates an electrical gradient used by the mitochondria to create ATP.

In addition to its enzymatic functions, copper is used in the electron transport chain. Interestingly, two copper-containing proteins that transport electrons, *azurin* and *plastocyanin*, have an intense blue color.

COPPER AND COLLAGEN

Copper is required by an enzyme, *lysyl oxidase*, which cross-links *collagen* and *elastin*. This cross-linking is essential for strong and flexible connective tissue, including skin and cartilage. Collagen is also a key element in the organic matrix of bone. Bones need copper for strength and density. Copper works with an enzyme to keep the connective tissue in the heart and blood vessels strong and flexible.

COPPER AND IRON TRANSPORT

Copper-containing enzymes are able to oxidize ferrous iron (Fe^{++}) to ferric iron (Fe^{3+}). This enables the iron to be loaded onto its transport protein, transferrin. Transferrin can transport iron to the bone marrow for red blood cell formation. Iron transportation is impaired in copper deficiency.

COPPER AND NEUROTRANSMITTERS

A copper-containing enzyme converts the neurotransmitter dopamine to the neurotransmitter norepinephrine. Copper-dependant enzymes are also involved in the metabolism of serotonin, a neurotransmitter that is well-known for influencing mood.

COPPER AND ANTIOXIDANT ACTIVITY

One of the most powerful antioxidants in the body is superoxide dismutase (SOD). Two forms of SOD contain copper. One is copper/zinc SOD that protects red blood cells. The other is an extracellular SOD found in high amounts in the lungs. SOD protects against the superoxide radical by converting it to hydrogen peroxide, which is then broken down to water. Copper also helps prevent iron from becoming a free radical.

Summary for Copper

Main functions: energy production, collagen synthesis, iron transport, and as an antioxidant.
RDA: adults, 900 mcg.
Toxicity: rare.
Tolerable upper intake level is 10 mg for adults.
Deficiency can cause anemia.
Sources: nuts and seeds, avocados, and green leafy vegetables such as spinach.
Forms in the body: found in cytochrome c oxidase, lysyl oxidase, and some forms of superoxide dismutase.

COPPER DEFICIENCY

Copper deficiency is rare. One of the clinical signs of copper deficiency is anemia that does not respond to iron supplementation. Copper deficiency can also lead to low numbers of white blood cells, which can increase susceptibility to infection. Premature and malnourished infants are vulnerable to copper deficiency if they are

fed formula made with cow's milk, which is low in copper. Children with cystic fibrosis may be more vulnerable to copper deficiency.

FOOD SOURCES OF COPPER

Copper is found in high amounts in nuts and seeds, avocados, and green leafy vegetables such as spinach, as seen in Graph 13-2. Whole grain products are also good sources of copper. Certain organ meats and shellfish are high in copper.

When copper is first absorbed in the intestines it is transported to the liver bound to albumin. Copper is carried in the bloodstream bound to a plasma protein called *ceruloplasmin*. Excess copper is removed from the liver into the bile. When the bile enters the intestines, the copper is given another chance at absorption.

Prolonged zinc supplementation above the recommended upper level of intake of 40 mg can reduce copper absorption. High levels of zinc increase intestinal production of *metallothionein*, which binds certain metals and can prevent their absorption. On the other hand, high iron intakes in infants may interfere with copper absorption.

RDAs have been established for copper to prevent deficiency, as seen in Table 13-3. These RDAs are designed to reduce the chance of copper deficiencies that may limit production of the important antioxidant superoxide dismutase.

Copper supplements come in several forms such as cupric oxide, copper gluconate, copper sulfate, and amino acid chelates of copper.

Table 13-3 RDAs and adequate intakes (AI) for copper for all ages.

RDAs for Copper	Age	Males (mcg/day)	Females (mcg/day)
Infants	0–6 months	200 (AI)	200 (AI)
Infants	7–12 months	220 (AI)	220 (AI)
Children	1–3 years	340	340
Children	4–8 years	440	440
Children	9–13 years	700	700
Adolescents	14–18 years	890	890
Adults	19 years and older	900	900
Pregnancy	all ages	—	1,000
Breastfeeding	all ages	—	1,300

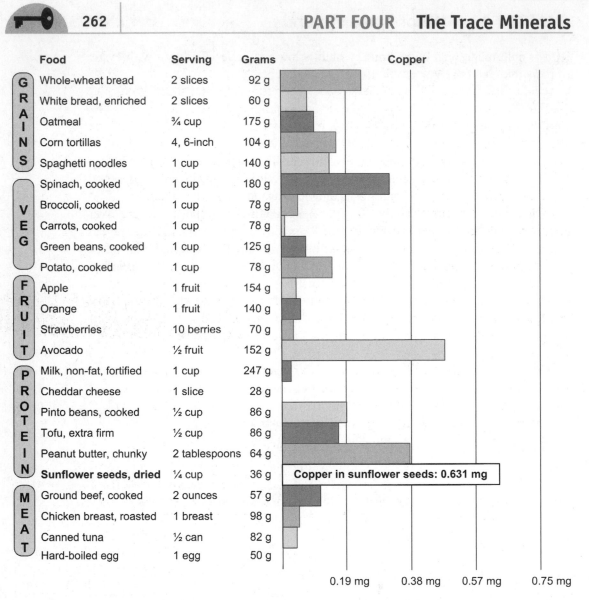

Food	Serving	Grams	Copper
GRAINS			
Whole-wheat bread	2 slices	92 g	
White bread, enriched	2 slices	60 g	
Oatmeal	¾ cup	175 g	
Corn tortillas	4, 6-inch	104 g	
Spaghetti noodles	1 cup	140 g	
VEG			
Spinach, cooked	1 cup	180 g	
Broccoli, cooked	1 cup	78 g	
Carrots, cooked	1 cup	78 g	
Green beans, cooked	1 cup	125 g	
Potato, cooked	1 cup	78 g	
FRUIT			
Apple	1 fruit	154 g	
Orange	1 fruit	140 g	
Strawberries	10 berries	70 g	
Avocado	½ fruit	152 g	
PROTEIN			
Milk, non-fat, fortified	1 cup	247 g	
Cheddar cheese	1 slice	28 g	
Pinto beans, cooked	½ cup	86 g	
Tofu, extra firm	½ cup	86 g	
Peanut butter, chunky	2 tablespoons	64 g	
Sunflower seeds, dried	¼ cup	36 g	Copper in sunflower seeds: 0.631 mg
MEAT			
Ground beef, cooked	2 ounces	57 g	
Chicken breast, roasted	1 breast	98 g	
Canned tuna	½ can	82 g	
Hard-boiled egg	1 egg	50 g	

0.19 mg   0.38 mg   0.57 mg   0.75 mg

Graph 13-2   Copper content of some common foods.

TOXICITY OF COPPER

Copper toxicity is rare. Occasional toxic levels of copper have resulted from drinking out of copper cups or from high levels in drinking water. The RDA for adults for copper is under one milligram. Just 30 milligrams of copper sulfate can be fatal. Copper in drinking water should be limited to well under two milligrams per liter. The tolerable upper intake level (UL) for children one to three years of age is 1 mg. For children four to eight years of age it is 3 mg. For older children the UL is 5 mg. For adolescents the UL is 8 mg and for adults the UL is 10 mg. Excess copper can inhibit the production of red blood cells. Long-term intake of copper over 10 milligrams daily has resulted in liver damage.

A rare inherited trait called *Wilson's disease* can result in an accumulation of copper in the body. Wilson's disease affects about one person in 30,000. The liver and the nervous system can become affected.

To summarize, copper has four important roles in the body. Copper helps with energy production, collagen synthesis, iron transport, and it catalyzes antioxidant activity. A little copper is needed, while too much copper is toxic.

Manganese

Manganese is essential in tiny quantities and is potentially toxic in larger amounts. The human body contains only about 12 milligrams of manganese. The chemical symbol for manganese is *Mn*. Manganese is a part of many enzyme systems in the body. Manganese assists an enzyme in the production of glucose when carbohydrates are not available. When excess amino acids are burned for energy, ammonia is produced in the body. A manganese-dependant enzyme detoxifies this ammonia.

ANTIOXIDANT ACTION OF MANGANESE

Manganese is especially important in protecting the mitochondria, the energy-producing organelles in the cell. As energy is produced in the mitochondria, powerful free radicals called *superoxide radicals* are formed. These superoxide radicals are converted to hydrogen peroxide by *manganese superoxide dismutase*, the principal antioxidant enzyme of mitochondria. The resulting hydrogen peroxide is reduced to water by other antioxidants.

MANGANESE AND COLLAGEN

Manganese participates in the formation of healthy bones and cartilage. Collagen in skin cells needs manganese in order to form. Manganese is needed by an enzyme that is needed for collagen formation in skin during wound healing.

MANGANESE DEFICIENCY

Manganese deficiency does not normally result in clear symptoms. Manganese deficiency may impair growth.

FOOD SOURCES OF MANGANESE

Whole grains, green leafy vegetables, and peanut butter are rich sources of manganese, as seen in Graph 13-3. White flour products are depleted of most of their manganese. Diets high in vegetables and whole grains are normally high in manganese.

There was not sufficient information for the Food and Nutrition Board of the Institute of Medicine to determine RDAs for manganese. Instead they published adequate intakes (AI), which are based on average consumption. Please refer to Table 13-4.

Table 13-4   Adequate intakes for manganese for all ages.

Adequate Intakes for Manganese	Age	Males mg/day	Females mg/day
Infants	0–6 months	0.003	0.003
Infants	7–12 months	0.6	0.6
Children	1–3 years	1.2	1.2
Children	4–8 years	1.5	1.5
Children	9–13 years	1.9	1.6
Adolescents	14–18 years	2.2	1.6
Adults	19 years and older	2.3	1.8
Pregnancy	all ages	—	2.0
Breastfeeding	all ages	—	2.6

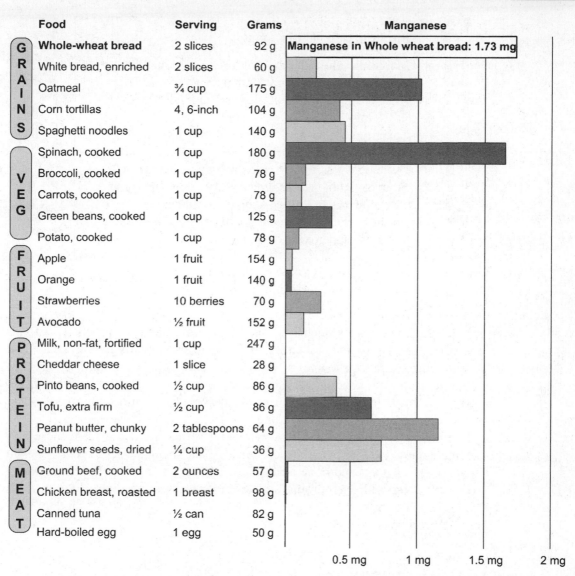

Food	Serving	Grams	Manganese
Whole-wheat bread	2 slices	92 g	Manganese in Whole wheat bread: 1.73 mg
White bread, enriched	2 slices	60 g	
Oatmeal	¾ cup	175 g	
Corn tortillas	4, 6-inch	104 g	
Spaghetti noodles	1 cup	140 g	
Spinach, cooked	1 cup	180 g	
Broccoli, cooked	1 cup	78 g	
Carrots, cooked	1 cup	78 g	
Green beans, cooked	1 cup	125 g	
Potato, cooked	1 cup	78 g	
Apple	1 fruit	154 g	
Orange	1 fruit	140 g	
Strawberries	10 berries	70 g	
Avocado	½ fruit	152 g	
Milk, non-fat, fortified	1 cup	247 g	
Cheddar cheese	1 slice	28 g	
Pinto beans, cooked	½ cup	86 g	
Tofu, extra firm	½ cup	86 g	
Peanut butter, chunky	2 tablespoons	64 g	
Sunflower seeds, dried	¼ cup	36 g	
Ground beef, cooked	2 ounces	57 g	
Chicken breast, roasted	1 breast	98 g	
Canned tuna	½ can	82 g	
Hard-boiled egg	1 egg	50 g	

GRAINS — VEG — FRUIT — PROTEIN — MEAT

0.5 mg   1 mg   1.5 mg   2 mg

Graph 13-3   Manganese content of some common foods.

Summary for Manganese

Main functions: glucose synthesis, ammonia detoxification, as an antioxidant, and in wound healing.

Adequate Intakes: adult men, 2.3 mg; adult women, 1.8 mg.

Toxicity: rare, no toxicity from food sources.

Tolerable upper intake level is 11 mg for adults.

Deficiency has few symptoms, may impair growth.

Sources: whole grains, green leafy vegetables, and peanut butter are rich sources of manganese.

Forms in the body: found in enzymes including manganese superoxide dismutase.

Foods high in phytates, such as beans, whole grains, and soy products, or foods high in oxalic acid, such as cabbage and spinach, may slightly inhibit manganese absorption. Teas are rich sources of manganese, although the tannins present in tea may reduce the absorption of manganese. Iron and manganese compete for absorption in the intestines. An excess of one will limit absorption of the other. Excess manganese is eliminated in bile.

Supplemental manganese comes in several forms including manganese sulfate, manganese gluconate, amino acid chelates, and manganese ascorbate. Special supplements for joint and bone health may contain large amounts of manganese ascorbate along with chondroitin sulfate and glucosamine hydrochloride. This type of supplement has been found to relieve pain in some cases of mild to moderate osteoarthritis of the knee.

MANGANESE TOXICITY

Manganese toxicity from food alone has not been reported, although some diets provide as much as 20 mg daily. The Food and Nutrition Board has set a rather conservative tolerable upper intake level (UL) of 11 mg for adults. The UL for children ranges from 2 mg to 6 mg and the UL for adolescents is 9 mg. High levels of manganese from drinking water may cause nerve problems similar to Parkinson's disease. The EPA recommends 0.05 milligrams per liter as the maximum allowable manganese concentration in drinking water.

To summarize, manganese is a lesser-known nutrient with many important functions. Manganese assists wound healing by helping to synthesize collagen. Manganese has great importance in activating an antioxidant in the mitochondria of the cells.

Fluoride

Fluoride is the ion form of the highly reactive element fluorine. The chemical symbol for fluorine is the letter F. Fluoride is not considered an essential trace mineral because it is not needed for growth or life. Almost all of the fluoride in the body is found in bones and teeth. Fluoride is best known for its role in preventing dental decay.

Bones and teeth contain calcium and phosphorus in a crystal structure known as *hydroxyapatite*. Fluoride can bond to the hydroxyapatite crystals, transforming the hydroxyapatite crystals into *fluoroapatite* crystals. The change from hydroxyapatite to fluoroapatite can harden tooth enamel during the years of tooth growth.

DEFICIENCY OF FLUORIDE

The only sign of deficiency of fluoride is an increased risk of tooth decay. In colder climates, the amount of fluoride in drinking water is 1.2 milligrams per liter. In warmer climates, where people drink more water, only 0.7 milligrams of fluoride per liter is used. Some studies have shown a reduced incidence of dental decay in populations drinking fluoridated water. There does not appear to be any advantage to fluoride supplements, topical fluoride, or fluoridated water in people over the age of 16, as the teeth have hardened by this age. In children and younger teens, topical fluoride treatments seem to reduce tooth decay.

INTAKE OF FLUORIDE

The major source of fluoride in the United States is fluoridated water. About two-thirds of the population of the United States drinks fluoridated water. The fluoride content of food is extremely low. Only tea and sardines have been found to contain even small amounts of fluoride. The swallowing of fluoride-containing toothpaste is a major source of fluoride intake, especially in children.

TOXICITY OF FLUORIDE

There is some controversy about the toxicity of fluoride. It is well established that excess fluoride can cause dental *fluorosis*. Mild dental fluorosis can result in mottled teeth. With higher levels of fluoride, pitting of the tooth enamel can occur. The Centers for Disease Control estimate that 23 percent of people in the United States aged 6 to 39 years of age have some degree of dental fluorosis. This may be due to the combined fluoride intake from fluoridated water, fluoridated toothpaste, and topical dental fluoride treatments.

Summary for Fluoride

Main functions: strengthens tooth enamel.
Adequate Intakes: adult men, 3.8 mg; adult women, 3.1 mg.
Toxicity: very toxic.
Tolerable upper intake level is 10 mg for adults.
Deficiency may increase risk of dental decay.
Sources: fluoridated drinking water.
Forms in the body: fluoroapatite in bones and teeth.

More research needs to be done to determine if normal levels of fluoride added to water cause lower intelligence in children or cause earlier onset of puberty in girls. High levels of fluoride are certainly toxic and can result in *skeletal fluorosis*, which can cause deformities.

To summarize, current information indicates that growing children can greatly reduce their risk of dental decay by receiving periodic fluoride applications to the surface of their teeth. Fluoride is an interesting trace mineral because of the differing viewpoints on its toxicity.

Chromium

Chromium was named after the Greek word for color (*chroma*) because of the colorful compounds made from it. The chemical symbol for chromium is *Cr*. One form of chromium, *trivalent chromium*, is essential for nutrition. The chemical symbol for trivalent chromium is Cr^{3+} (Cr^{+++}). Trivalent chromium is the form used in the body and the form found in food. A different form of chromium is *hexavalent chromium* (Cr^{6+}). Hexavalent chromium is used in industry, is irritating, and causes cancer.

CHROMIUM AND BLOOD SUGAR

Trivalent chromium enhances the effects of *insulin*. The body secretes the hormone insulin in response to rising levels of blood sugar. Insulin binds to receptors on cell membranes thus stimulating the cells to take in more glucose (blood sugar). This not only clears excess insulin from the blood, but also assists the cells in obtaining blood sugar. Compounds that assist insulin in clearing glucose from the blood are called *glucose tolerance factors*. Some glucose tolerance factors contain chromium. Trivalent chromium also enhances the ability of insulin to remove fats from the blood.

A decreased response to insulin can result in *impaired glucose tolerance*. A more serious insulin resistance is known as *type 2 diabetes*. The clinical signs of type 2 diabetes include elevated blood sugar levels and insulin resistance.

Cell membranes have insulin receptors. The sensitivity of this receptor to insulin can be improved by chromium. Insulin binds to the insulin receptor in the cell membrane to activate the receptor, as seen in Figure 13-4. The activation of the insulin receptor enables glucose and chromium to enter the cell. Chromium binds to the insulin receptor and enhances its activity. With available chromium, more glucose enters the cell. This is how chromium availability helps to relieve impaired glucose tolerance and type 2 diabetes.

CHROMIUM DEFICIENCY

Chromium deficiency is difficult to determine because of the lack of accurate tests for chromium status. Chromium deficiency may be a contributing factor in both type 2 diabetes and impaired glucose tolerance. Heavy exercise may increase the amount of chromium needed.

Summary for Chromium

Main function: assists insulin in controlling blood sugar.
Adequate Intakes: adults, 20 mcg to 35 mcg.
Toxicity: no reports of toxicity from trivalent chromium.
Hexavalent chromium is highly toxic.
Tolerable upper intake level has not been set.
Deficiency may interfere with the control of blood sugar.
Sources: whole grain products, broccoli, green beans,
    grape juice, and spices.
Form in the body: trivalent chromium.

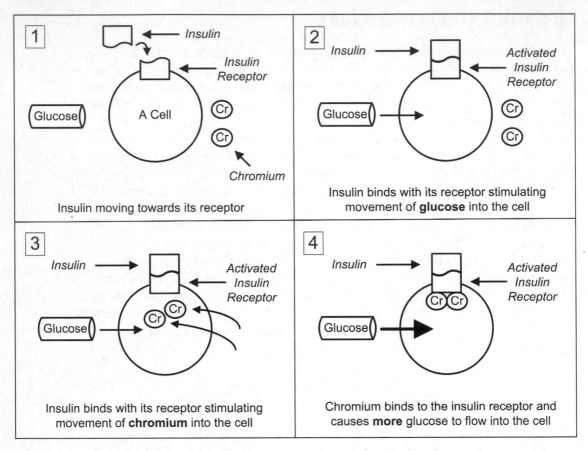

Figure 13-4   Chromium increases the power of insulin.

FOOD SOURCES FOR CHROMIUM

Data on the chromium content of foods is limited. Furthermore, the amount of chromium in foods is quite variable. Good sources of chromium include whole grain products, broccoli, green beans, grape juice, and spices. Processed foods and foods high in sugar are poor sources of chromium. Foods high in sucrose and fructose increase chromium loss. Vitamin C in amounts of 100 mg or more increase the absorption of chromium. There is not enough information on chromium to establish RDAs. The adequate intake level (AI) has been set to reflect average intakes. Please refer to Table 13-5 for the adequate intake levels for chromium.

CHROMIUM SUPPLEMENTS

Trivalent chromium is available for supplementation in several forms. Chromium chloride is a poorly absorbed form. Chromium picolinate is the form used in much of the research on the health effects of chromium and is well absorbed. Chromium nicotinate also has good bioavailability. Certain nutritional yeast products are fortified with chromium and have good bioavailability.

TOXICITY OF CHROMIUM

As noted above, hexavalent chromium is known to cause cancer and skin irritation. There have been no convincing reports of toxicity from the trivalent chromium in food or supplements. Therefore, the Food and Nutrition Board has not set an upper level for chromium. Several studies have shown a lack of side effects from supplementation of 1000 mcg for several months. People with kidney or liver disease should be cautioned to take lower doses. Normal supplementation levels are between 60 mcg and 120 mcg of chromium.

Chromium is a valuable trace mineral because it boosts the power of insulin.

Table 13-5   Adequate intakes for chromium for all ages.

Adequate Intake for Chromium	Age	Males (mcg/day)	Females (mcg/day)
Infants	0–6 months	0.2	0.2
Infants	7–12 months	5.5	5.5
Children	1–3 years	11	11
Children	4–8 years	15	15
Children	9–13 years	25	21
Adolescents	14–18 years	35	24
Adults	19–50 years	35	25
Adults	51 years and older	30	20
Pregnancy	18 years and younger	—	29
Pregnancy	19 years and older	—	30
Breastfeeding	18 years and younger	—	44
Breastfeeding	19 years and older	—	45

Molybdenum

Molybdenum is an essential trace mineral for virtually all life forms on Earth. The chemical symbol for molybdenum is *Mo*. The name molybdenum comes from the Greek word *molybdos*, which means lead-like. Molybdenum is used in the body as a cofactor in three enzymes.

The most important enzyme that uses molybdenum is *sulfite oxidase*. Sulfite oxidase is needed to add oxygen to transform sulfite (SO_3^{2-}) to sulfate (SO_4^{2-}). This transformation is necessary for the metabolism of the sulfur-containing amino acids cysteine and methionine.

Molybdenum is used in another important enzyme called *xanthine oxidase*. Xanthine oxidase contains two molybdenum atoms in addition to eight iron atoms and riboflavin. Xanthine oxidase breaks down parts of DNA into uric acid in the bloodstream. The third enzyme to use molybdenum is *aldehyde oxidase*. Aldehyde oxidase works with xanthine oxidase to assist in the metabolism of drugs and toxins.

Molybdenum is needed in the soil used for food crops. If plants do not get enough molybdenum they may contain more cancer-causing nitrosamines. With sufficient molybdenum, plants can convert nitrates to amino acids, lowering the nitrosamine levels.

Summary for Molybdenum

Main function: metabolism of sulfur-containing amino acids.

RDA: adults, 45 mcg.

Toxicity: low toxicity.

Tolerable upper intake level has been set at 2 mg.

Deficiency is rare.

Sources: beans, lentils, whole grains, and nuts.

Forms in the body: found in the enzymes sulfite oxidase, xanthine oxidase, and aldehyde oxidase.

DEFICIENCY OF MOLYBDENUM

Molybdenum deficiency has not been observed in healthy people. There are rare cases of molybdenum deficiency from genetic defects or from faulty intravenous

feeding. Signs of molybdenum deficiency may include low levels of uric acid in the blood and urine and an increase of sulfite in the urine.

DIETARY SOURCES OF MOLYBDENUM

A survey of molybdenum intake of Americans indicated that average intakes are above the RDA. The richest dietary sources for molybdenum are beans, lentils, and peas. Whole grains and nuts are good sources. Other foods, such as animal products, vegetables, and fruits are low in molybdenum. The molybdenum content of food can vary considerably.

The Food and Nutrition Board of the Institute of Medicine has set RDAs for molybdenum, as seen in Table 13-6. Adequate Intakes (AI) were set for infants.

Molybdenum found in nutritional supplements is often in the form of *sodium molybdate* or *ammonium molybdate*.

TOXICITY OF MOLYBDENUM

Molybdenum has a low toxicity. Exposure to very high intakes of 10 mg to 15 mg has resulted in increased levels of uric acid in the blood and gout-like symptoms. The tolerable upper intake level has been set at a conservative 2 mg for adults.

To summarize, molybdenum is needed in trace amounts by enzymes that metabolize the sulfur-containing amino acids. It also assists in the removal of toxins from the blood.

Table 13-6   RDAs and adequate intakes (AI) for molybdenum for all ages.

RDAs for Molybdenum	Age	Males mcg/day	Females mcg/day
Infants	0–6 months	2 (AI)	2 (AI)
Infants	7–12 months	3 (AI)	3 (AI)
Children	1–3 years	17	17
Children	4–8 years	22	22
Children	9–13 years	34	34
Adolescents	14–18 years	43	43
Adults	19 years and older	45	45
Pregnancy	all ages	—	50
Breastfeeding	all ages	—	50

Other Trace Minerals

Some trace minerals may be needed in tiny amounts. However, in large amounts, they can be toxic. If supplementation is desired because of lack of trace minerals in food grown in deficient soil, it is recommended that the supplement have a balanced amount of each trace mineral. Supplementation with single trace minerals can throw off mineral balances and can be toxic in larger amounts.

NICKEL

Nickel (Ni) is an essential trace mineral. Nickel is needed by certain enzymes used in anaerobic energy production in the cell. Nickel works with iron and sulfur to release energy from carbohydrates.

SILICON

The human body contains about 35 grams of silicon. Silicon (Si) strengthens connective tissue such as bones, cartilage, blood vessels, and tendons. Silicon is found in whole grains and fresh vegetables.

VANADIUM

Vanadium (V) is named after the Norse goddess of love and beauty. Vanadium plays a role in bone growth.

COBALT

Cobalt (Co) is the central atom in the vitamin B_{12} molecule. While vitamin B_{12} is an essential nutrient, cobalt has not been established as an essential nutrient.

BORON

Boron (B) has been found to be important for mental acuity and brain function. Boron may also be important for the functioning of membranes. With further research boron may become recognized as an essential trace mineral.

Toxic Heavy Metals

Lead and mercury are toxic heavy metals that should be avoided.

LEAD

The chemical symbol for lead, *Pb*, comes from the Latin root word for plumbing. Lead is similar to some other ions with two positive charges such as calcium (Ca^{++}) and iron (Fe^{++}). Lead can take the place of iron or calcium, but lead cannot perform the functions that iron or calcium can perform. For instance, lead can take the place of iron in heme in hemoglobin, but this hemoglobin can no longer carry oxygen. Lead can take the place of calcium in brain cells, but the brain cells are no longer capable of processing messages from nerve cells.

Children who are exposed to lead are more likely to have lower intelligence quotients and to develop learning disabilities and behavioral problems than normal children. If women are exposed to lead during pregnancy, especially if they are calcium-deficient, their children may have abnormal neurological development. Lead exposure in adults can increase the risk of kidney damage and high blood pressure.

Lead can also be incorporated into bones. This lead can remain in the bones for decades. During pregnancy, if insufficient dietary calcium is taken, calcium and lead may be removed from bones and put into the blood circulation with damaging effects on the fetus. Lead can easily pass through the placenta and damage the vulnerable nervous system of the growing fetus. This is one reason for mothers to have sufficient daily intake of calcium, which makes the removal of calcium (and possibly lead) from bones unnecessary. During postmenopausal years, women can reduce their lead levels by taking in sufficient calcium so that bone demineralization does not occur.

> **Three Ways to Lower Blood Lead Levels**
>
> Adequate dietary calcium
> Adequate dietary iron
> Supplementary vitamin C

Sufficient dietary calcium can reduce the absorption of lead in the intestines. Another way to reduce lead levels is to have adequate dietary iron. Iron deficiency can increase the blood levels of lead in children. A third way to reduce levels of lead in blood is to take in higher levels of vitamin C. Vitamin C intakes of 100 mg to 1000 mg daily have been associated with lower blood levels of lead. These protective levels of vitamin C are difficult to obtain from diet alone.

Lead has many ill effects on the body. Lead can make red blood cells leaky and fragile. Lead impairs the immune system in two ways. Lead interferes with the ability of white blood cells to fight infection. Lead can also bind to antibodies and impair their effectiveness. Lead can interfere with growth and tooth development. Lead exposure should be minimized, especially in children, because of its many harmful effects.

MERCURY

Mercury is a toxic heavy metal. The chemical symbol for mercury is *Hg*, which comes from the Greek word for quicksilver. Mercury exposure is most damaging to developing fetuses and nursing babies, where it can cause severe problems with nerve development. Mercury consumption is also damaging to children and adults. Mercury damages the central nervous system, endocrine system, kidneys, and adversely affects the mouth, gums, and teeth. Most of the exposure to mercury in America is from eating fish and shellfish. Fish and shellfish accumulate a highly toxic organic form of mercury called *methylmercury*.

Quiz

Refer to the text in this chapter if necessary. A good score is at least 8 correct answers out of these 10 questions. The answers are listed in the back of this book.

1. Which form is found in food?
   (a) Iodine.
   (b) Iodide.
   (c) Thyroxine.
   (d) Triiodothyronine.

2. Iodide, in thyroid hormones, controls:
   (a) Growth.
   (b) Reproduction.
   (c) Metabolism.
   (d) All of the above.

3. Iodine deficiency during pregnancy can cause:
   (a) Beriberi.
   (b) Cretinism.
   (c) Pellagra.
   (d) The flu.

4. Selenium works in glutathione peroxidase to convert:
   (a) Hydrogen peroxide to water.
   (b) Superoxide to hydrogen peroxide.
   (c) Thyroxine to triiodothyronine.
   (d) Glucose to water.

5. Selenium is found in high levels in:
   (a) Broccoli.
   (b) Green beans.
   (c) Brazil nuts.
   (d) Strawberries.

6. Copper protects red blood cells against oxidation in:
   (a) Vitamin C.
   (b) Vitamin E.
   (c) Glutathione peroxidase.
   (d) Superoxide dismutase.

7. Manganese toxicity is:
   (a) Rare.
   (b) Somewhat common.
   (c) Common.
   (d) A widespread problem.

8. Which mineral helps reduce dental decay?
   (a) Zinc.
   (b) Fluoride.
   (c) Manganese.
   (d) Chromium.

9. Chromium enhances the effects of:

(a) Serotonin.

(b) Insulin.

(c) Adrenaline.

(d) Thyroxine.

10. The healthiest amount of mercury to intake daily in food is:

(a) 10 mg.

(b) 10 mcg.

(c) 1 mcg.

(d) 0 mcg.

Test: Part Four

Do not refer to the text when taking this test. A good score is at least 18 (out of 25 questions) correct. Answers are in the back of the book. It's best to have a friend check your score the first time, so that you won't memorize the answers if you want to take the test again.

1. Iron deficiency affects about how many people worldwide?
    (a) One hundred thousand.
    (b) One million.
    (c) One hundred million.
    (d) One billion.

2. Iron participates in which part of energy metabolism?
    (a) Pyruvic acid synthesis.
    (b) Amino acid conversion.
    (c) The electron transport chain.
    (d) Anaerobic energy production.

3. Hemoglobin:

   (a) Transports oxygen in blood.

   (b) Contains about two-thirds of the iron in the body.

   (c) Releases oxygen to tissues.

   (d) All of the above.

4. An iron-containing compound in the electron transport chain:

   (a) Cytochrome.

   (b) Hydrogen peroxide.

   (c) Myoglobin.

   (d) Hemoglobin.

5. Copper and vitamin A:

   (a) Worsen iron deficiency.

   (b) Help relieve iron deficiency.

   (c) Have no effect on iron deficiency.

   (d) Vitamin A relieves iron deficiency, but copper worsens iron deficiency.

6. Dietary iron is needed most:

   (a) In children under four years of age.

   (b) In early adolescence.

   (c) During pregnancy.

   (d) All of the above.

7. Iron is transported by:

   (a) Transferrin.

   (b) Mitochondria.

   (c) Hemosiderin.

   (d) Metalloenzymes.

8. Iron absorption is increased by:

   (a) Phytates.

   (b) Oxalates.

   (c) Vitamin C.

   (d) Tannic acid.

9. In zinc finger-like proteins, zinc attaches to:
   (a) Cysteine.
   (b) Histidine.
   (c) Both (a) and (b).
   (d) Neither (a) nor (b).

10. Zinc stabilizes the antioxidant:
   (a) Superoxide dismutase.
   (b) Action in the mitochondria.
   (c) Glutathione peroxidase.
   (d) Vitamin E.

11. Severe zinc deficiency:
   (a) Is caused by insufficient dietary intake.
   (b) Is rare.
   (c) Cannot be caused by diarrhea.
   (d) Is not caused by severe burns.

12. Which is the richest source of zinc?
   (a) Fruits and vegetables.
   (b) Beans.
   (c) Oysters.
   (d) Enriched white bread.

13. Zinc is bound in the intestines and in the liver by:
   (a) Transferrin.
   (b) Metallothionein.
   (c) Ferrin.
   (d) Hemosiderin.

14. The RDA for zinc for adult men is:
   (a) 11 mg.
   (b) 11 mcg.
   (c) 1.1 mg.
   (d) 1.1 mcg.

15. Excess zinc:

    (a) Can cause a copper deficiency.

    (b) Can reduce the effectiveness of superoxide dismutase.

    (c) Can result from the use of zinc lozenges.

    (d) All of the above.

16. The most active hormone made from iodine is:

    (a) Adrenaline.

    (b) Thyroxine (T4).

    (c) Triiodothyronine (T3).

    (d) Serotonin.

17. Which gland(s) control thyroid hormone production?

    (a) Hypothalamus.

    (b) Pituitary.

    (c) Both (a) and (b).

    (d) Neither (a) nor (b).

18. Which form of selenium supplementation is most absorbable?

    (a) Sodium selenite.

    (b) Sodium selenate.

    (c) Selenomethionine.

    (d) Selenium oxide.

19. Copper is used in which antioxidant?

    (a) Superoxide dismutase.

    (b) Cytochrome.

    (c) Vitamin C.

    (d) Glutathione peroxidase.

20. Excess copper:

    (a) Is eliminated by the kidneys.

    (b) Is eliminated in the bile.

    (c) Is eliminated by intestinal cells.

    (d) Is lost in sweat.

21. A manganese-dependent enzyme detoxifies:
    (a) Acetone.
    (b) Drugs.
    (c) Ammonia.
    (d) Bile.

22. Rich manganese sources include all of the following EXCEPT:
    (a) Whole grains.
    (b) Hamburger.
    (c) Green leafy vegetables.
    (d) Peanut butter.

23. Manganese toxicity from food:
    (a) Is very rare.
    (b) Is common.
    (c) Is a worldwide health problem.
    (d) Is only a problem in populations that eat oysters.

24. When fluoride is incorporated into bones, it can form:
    (a) Thyroxine.
    (b) Fluorine.
    (c) Hydroxyapatite.
    (d) Fluoroapatite.

25. Trace minerals are found in the body in:
    (a) Large amounts.
    (b) Medium amounts.
    (c) Small amounts.
    (d) Trace amounts.

Final Exam

Do not refer to the text when taking this exam. A good score is at least 75 (out of 100 questions) correct. Answers are in the back of the book. It's best to have a friend check your score the first time, so that you won't memorize the answers if you want to take the test again.

1. Water-soluble vitamins:
   - (a) Are first absorbed into the lymph fluid.
   - (b) Are absorbed directly into blood.
   - (c) Are found only in fruits and vegetables.
   - (d) All of the above.

2. Thiamin is known as:
   - (a) Vitamin B_1.
   - (b) Vitamin B_2.
   - (c) Vitamin B_3.
   - (d) Vitamin B_4.

3. Abundant natural B vitamins are found in:

   (a) B-complex vitamin supplements.

   (b) Soybean oil.

   (c) Peaches.

   (d) Nutritional yeast.

4. The form of vitamin B_3 that can cause skin flushing is:

   (a) Niacinamide.

   (b) Niacin.

   (c) Nicotinamide.

   (d) All of the above.

5. Riboflavin helps reactivate this important antioxidant:

   (a) Tryptophan.

   (b) Cobalamin.

   (c) Biotin.

   (d) Glutathione.

6. Part of our niacin requirement can be met by dietary:

   (a) Tryptophan.

   (b) Lysine.

   (c) Methionine.

   (d) Leucine.

7. Niacin deficiency can cause:

   (a) Blindness.

   (b) Pellagra.

   (c) Beriberi.

   (d) Scurvy.

8. Which B vitamin makes up a large part of coenzyme A?

   (a) Pantothenic acid.

   (b) Tryptophan.

   (c) Thiamin.

   (d) Biotin.

9. Coenzyme A is an important part of:

   (a) Energy production.

   (b) The TCA cycle.

   (c) The Krebs cycle.

   (d) All of the above.

10. When white flour is made from whole grains, what percentage of pantothenic acid is lost?

    (a) 43 percent.

    (b) 28 percent.

    (c) 12 percent.

    (d) 5 percent.

11. Which vitamin is extensively stored in muscle tissue?

    (a) Niacin.

    (b) Vitamin C.

    (c) Pyridoxine.

    (d) Vitamin B_{12}.

12. Folic acid, as opposed to folates, is found in:

    (a) Food.

    (b) Vitamin pills.

    (c) The body.

    (d) A metabolically active form in the body.

13. Which of the following is active in eliminating homocysteine from blood?

    (a) Pyridoxine.

    (b) Folic acid.

    (c) Vitamin B_{12}.

    (d) All of the above.

14. Excess folate is removed from the body:

    (a) Through the lungs.

    (b) In urine.

    (c) In bile.

    (d) None of the above.

15. Supplementary vitamin B_{12} is often in the form of:
    (a) Methylcobalamin.
    (b) Deoxyadenosyl cobalamin.
    (c) Cyanocobalamin.
    (d) Magentocobalamin.

16. Which has the highest energy level?
    (a) ADP.
    (b) AMP.
    (c) APP.
    (d) ATP.

17. Which one is a real vitamin?
    (a) Vitamin K.
    (b) Vitamin T.
    (c) Vitamin P.
    (d) Vitamin U.

18. Which of the following is NOT a cure for scurvy?
    (a) The tips of young arbor vitae evergreen needles.
    (b) Sauerkraut.
    (c) Limes.
    (d) Fresh fish.

19. Vitamin C is:
    (a) Synthesized by only a few animals.
    (b) Synthesized by humans.
    (c) Synthesized by most animals.
    (d) Must be obtained from food for all animals.

20. The amount of vitamin C synthesized by a 154-pound dog each day is:
    (a) 2800 mg.
    (b) 280 mg.
    (c) 28 mg.
    (d) 2.8 mg.

21. Collagen is a major structural component of:
    (a) Blood vessel walls.
    (b) Skin.
    (c) Teeth.
    (d) All of the above.

22. Which is NOT needed to synthesize collagen?
    (a) Lysine.
    (b) Magnesium.
    (c) Iron.
    (d) Proline.

23. The difference between ascorbic acid and dehydroascorbic acid is:
    (a) Only ascorbic acid comes from food.
    (b) Dehydroascorbic acid is the oxidized form of ascorbic acid.
    (c) Ascorbic acid is the oxidized form of dehydroascorbic acid.
    (d) Dehydroascorbic acid is found only in secret government labs.

24. Vitamin C:
    (a) Prevents colds.
    (b) Has no effect on colds.
    (c) Lessens the severity of colds.
    (d) Lengthens the duration of colds.

25. Ascorbated vitamin C:
    (a) Does not have a neutral pH.
    (b) Is ascorbic acid attached to a mineral.
    (c) Always contains bioflavonoids.
    (d) Creates intestinal irritation in large doses.

26. Which form of vitamin A is not called "preformed"?
    (a) Retinyl palmitate.
    (b) Beta-carotene.
    (c) Retinal.
    (d) Retinol.

27. The form of Vitamin A that is an antioxidant is:
    (a) Retinol.
    (b) Retinyl esters.
    (c) Retinoic acid.
    (d) None of the above.

28. The form of vitamin A that is used for transport of the vitamin in the body is:
    (a) Retinol.
    (b) Retinyl esters.
    (c) Retinoic acid.
    (d) None of the above.

29. Beta-carotene is abundant in:
    (a) Dairy products, especially milk.
    (b) Yellow and orange vegetables and fruit.
    (c) Fish, especially salmon.
    (d) Beef liver.

30. Vitamin A reduces risk of infections because:
    (a) It strengthens mucous membranes.
    (b) It is needed in the development of lymphocytes.
    (c) It is needed for the regulation of the immune system.
    (d) All of the above.

31. Vitamin A in the form of retinoic acid:
    (a) Enables red blood cell production.
    (b) Enables fetal development.
    (c) Enables cell differentiation.
    (d) All of the above.

32. The type of vitamin A NOT associated with toxicity is:
    (a) Beta-carotene supplements.
    (b) Beta-carotene in food.
    (c) Vitamin A in liver.
    (d) Supplements of vitamin A.

33. To prevent osteoporosis in older people:
    (a) Keep vitamin A in the diet and supplements at a minimum.
    (b) Eat the RDA of vitamin A in food.
    (c) Take double the RDA of vitamin A in supplement form.
    (d) Take vitamin A only in the form of retinoic acid.

34. Calcidiol is:
    (a) Stored in the liver.
    (b) The active form of vitamin D.
    (c) Made in the skin from sunlight.
    (d) Made by irradiating a fungus.

35. People in areas far from the equator may get too little vitamin D because:
    (a) The days are shorter.
    (b) Protective clothing is common.
    (c) The angle of the sun decreases UVB penetration through the atmosphere.
    (d) All of the above.

36. Vitamin D, in the form of calcitriol:
    (a) Regulates calcium and magnesium levels in the blood.
    (b) Regulates phosphorus and magnesium levels in the blood.
    (c) Regulates calcium and phosphorus levels in the blood.
    (d) Regulates calcium and sodium levels in the blood.

37. Blood calcium levels can be increased:
    (a) By increased absorption of calcium from the intestines.
    (b) By decreasing kidney losses of calcium.
    (c) By releasing stored calcium from bones.
    (d) All of the above.

38. Severe lack of vitamin D can cause:
    (a) Rickets.
    (b) Pellagra.
    (c) Scurvy.
    (d) Diphtheria.

39. Food sources of vitamin D include:
    (a) Fruits and vegetables.
    (b) Salmon and mackerel.
    (c) Soybean oil.
    (d) Pizza.

40. Gamma-tocopherol:
    (a) Is the preferred form of vitamin E for protecting LDL.
    (b) Is the most abundant form of vitamin E in the blood.
    (c) Is the form of vitamin E found most abundantly in the American diet.
    (d) Is the form of vitamin E that absorbs easily through skin.

41. When vitamin E becomes oxidized, it can be reactivated by:
    (a) The ascorbate form of vitamin C.
    (b) Vitamin A.
    (c) The kidneys.
    (d) Vitamin D.

42. Vitamin E:
    (a) Thins blood by reducing platelet clumping.
    (b) Makes arteries more flexible.
    (c) Reduces oxidation of LDL.
    (d) All of the above.

43. The average American diet provides:
    (a) An abundance of vitamin E.
    (b) Just enough vitamin E to prevent heart disease.
    (c) Less than optimal amounts of vitamin E.
    (d) Plenty of vitamin E if enough fruits and vegetables are eaten.

44. Natural vitamin E is:
    (a) dl-alpha-tocopherol.
    (b) Mixed tocopherols from food.
    (c) SRR-alpha-tocopherol.
    (d) All of the above.

45. Enriched white flour, compared to whole wheat flour, has:
    (a) No vitamin E.
    (b) Two percent of the vitamin E.
    (c) The same amount of vitamin E.
    (d) Twice as much vitamin E.

46. Vitamin K is:
    (a) Water soluble.
    (b) Fat soluble.
    (c) Not soluble.
    (d) Mineral soluble.

47. Vitamin K converts amino acid residues in certain proteins so that:
    (a) The proteins can bind magnesium.
    (b) The proteins can bind phosphorus.
    (c) The proteins can bind sodium.
    (d) The proteins can bind calcium.

48. Vitamin K is needed to synthesize two proteins, proteins C and S, that:
    (a) Accelerate clotting.
    (b) Slow clotting.
    (c) Are unrelated to clotting.
    (d) Are needed for strong bones.

49. The Dietary Reference Intake (DRI) of vitamin K for an adult man is:
    (a) 90 micrograms.
    (b) 90 milligrams.
    (c) 120 micrograms.
    (d) 120 milligrams.

50. Breast-fed newborns may be deficient in vitamin K because:
    (a) Breast milk is low in vitamin K.
    (b) Intestinal bacteria may not be present to make vitamin K.
    (c) The vitamin K conservation cycle may not be operational.
    (d) All of the above.

51. Water intake must:
    (a) Exactly match water output.
    (b) Exceed water output.
    (c) Be less than water output.
    (d) Be one gallon more than water output.

52. The body loses about 33 percent of its water through:
    (a) Sweating.
    (b) Skin diffusion.
    (c) Sweating plus skin diffusion.
    (d) The feces.

53. Which beverage does NOT cause diuretic loss of water?
    (a) Coffee.
    (b) Water.
    (c) Beer.
    (d) Tea.

54. Which area of the body contains electrolytes?
    (a) Blood plasma.
    (b) Intracellular fluid.
    (c) Interstitial fluid.
    (d) All of the above.

55. The following means "inside the cell":
    (a) Intracellular.
    (b) Interstitial.
    (c) Extracellular.
    (d) Blood.

56. Renin:
    (a) Causes blood pressure to drop.
    (b) Causes the kidneys to retain less sodium.
    (c) Causes the kidneys to retain more sodium.
    (d) Causes the kidneys to retain less water.

57. Table salt dissolves into:
    (a) Sodium and potassium.
    (b) Sodium and chloride.
    (c) Sodium and magnesium.
    (d) Chloride and potassium.

58. Inside the cell, which are the most common electrolytes?
    (a) Potassium and phosphate.
    (b) Sodium and chloride.
    (c) Sodium and magnesium.
    (d) Chloride and potassium.

59. The sodium-potassium pump:
    (a) Moves potassium out of the cell.
    (b) Moves sodium out of the cell.
    (c) Moves sodium out of the plasma.
    (d) Moves sodium out of the extracellular fluid.

60. Electrolyte balance can get thrown off balance:
    (a) By excess sweating.
    (b) By prolonged vomiting or diarrhea.
    (c) By wounds.
    (d) All of the above.

61. This mineral is a macro mineral:
    (a) Iron.
    (b) Zinc.
    (c) Calcium.
    (d) Copper.

62. During food processing:
    (a) Sodium is reduced and potassium is increased.
    (b) Potassium is reduced and sodium is increased.
    (c) The sodium and potassium ratio is not changed much.
    (d) Both sodium and potassium are increased.

63. Hydrochloric acid in the stomach is made with:
    (a) Chloride.
    (b) Sodium.
    (c) Potassium.
    (d) Magnesium.

64. Which food is low in potassium?
    (a) Lettuce.
    (b) Carrots.
    (c) Steak.
    (d) Potatoes.

65. Hydroxyapatite in bones is made of:
    (a) Calcium and phosphorus.
    (b) Calcium and sodium.
    (c) Calcium and potassium.
    (d) Sodium and potassium.

66. Most of the calcium in the body is in the:
    (a) Blood.
    (b) Bones.
    (c) Brain.
    (d) Soft tissue.

67. Calcium channels open in cell membranes to:
    (a) Allow calcium to enter the cell.
    (b) Allow calcium to leave the cell.
    (c) Block calcium from passing.
    (d) Keep the calcium out of the cell.

68. Which organ controls calcium levels in the blood?
    (a) Brain.
    (b) Heart.
    (c) Lungs.
    (d) Parathyroid gland.

69. Low blood levels of calcium can be caused by:
    (a) Lack of vitamin A.
    (b) Lack of vitamin E.
    (c) Lack of vitamin D.
    (d) Lack of vitamin C.

70. The average American man needs to eat how much additional calcium daily to compensate for the average excesses of sodium and protein?
    (a) 100 mg.
    (b) 250 mg.
    (c) 350 mg.
    (d) Over 500 mg.

71. Which food is highest in calcium?
    (a) Milk.
    (b) Sesame seeds.
    (c) Hamburger.
    (d) Kale.

72. Most of the phosphorus in the body is in:
    (a) The bones.
    (b) The skin.
    (c) The hair.
    (d) The blood.

73. There are more phosphate groups in:
    (a) AMP.
    (b) ADP.
    (c) ATP.
    (d) Creatine phosphate.

74. Magnesium in the body is found mostly in the bones and:
    (a) Skin.
    (b) Blood.
    (c) Muscle.
    (d) Extracellular fluid.

75. Magnesium:

    (a) Relaxes muscles.

    (b) Reduces high blood pressure.

    (c) Increases calcium absorption into bones.

    (d) All of the above.

76. The chemical symbol for iron is:

    (a) Hg.

    (b) Mn.

    (c) Fe.

    (d) Zn.

77. Iron is in the center of:

    (a) Heme.

    (b) Chlorophyll.

    (c) White blood cells.

    (d) Pyruvic acid.

78. Myoglobin:

    (a) Is found in blood.

    (b) Is found in muscle.

    (c) Acquires oxygen from the lungs.

    (d) All of the above.

79. The body lowers blood levels of iron:

    (a) During infections.

    (b) After a meal with a large serving of spinach.

    (c) During exercise.

    (d) In summer months.

80. Iron:

    (a) Is not stored in the body.

    (b) Is stored for long periods of time.

    (c) Dietary lack appears as iron deficiency anemia immediately.

    (d) Is needed daily in the diet.

81. During pregnancy, iron is most needed:
    (a) In the prenatal period.
    (b) In the first trimester.
    (c) In the second trimester.
    (d) In the third trimester.

82. The amount of iron in the body is controlled by:
    (a) The kidneys.
    (b) The bile.
    (c) Varying absorption rates.
    (d) Sweating.

83. Zinc is important for:
    (a) Synthesis of insulin.
    (b) Storage of insulin.
    (c) Release of insulin.
    (d) All of the above.

84. Zinc helps stabilize finger-like proteins by:
    (a) Helping the proteins fold.
    (b) Working as a cofactor to enzymes.
    (c) Interfering with copper absorption.
    (d) Binding to arginine and lysine.

85. Large amounts of zinc can reduce the availability of:
    (a) Calcium.
    (b) Phosphorus.
    (c) Copper.
    (d) Sodium.

86. The best source of zinc for younger infants is:
    (a) Formulas based upon cow's milk.
    (b) Breast milk.
    (c) Formulas based upon soy milk.
    (d) Low-fat cow's milk.

87. What percentage of pregnant women worldwide have insufficient zinc levels?

    (a) 80 percent.

    (b) 60 percent.

    (c) 40 percent.

    (d) 20 percent.

88. Populations most susceptible to zinc deficiency include:

    (a) Children and infants.

    (b) Pregnant women.

    (c) Alcoholics.

    (d) All of the above.

89. Whole wheat flour has:

    (a) Half as much zinc as enriched flour.

    (b) The same amount of zinc as enriched flour.

    (c) Four times as much zinc as enriched flour.

    (d) Eight times as much zinc as enriched flour.

90. Zinc can return to the intestines in:

    (a) Pancreatic juice.

    (b) Bile.

    (c) Insulin.

    (d) Transferrin.

91. Iodine is needed by:

    (a) The hypothalamus gland.

    (b) The thyroid gland.

    (c) The liver.

    (d) The adrenal glands.

92. Deficiency of iodine can cause:

    (a) Beriberi.

    (b) Goiter.

    (c) Pellagra.

    (d) All of the above.

93. Which is NOT a good source of iodine?

    (a) Seaweed.

    (b) Fish.

    (c) Apples.

    (d) Iodized salt.

94. Selenium is part of which important antioxidant enzyme?

    (a) Superoxide dismutase.

    (b) Cytochrome.

    (c) Vitamin C.

    (d) Glutathione peroxidase.

95. Manganese is used as an antioxidant in the:

    (a) Mitochondria.

    (b) Blood.

    (c) Cell membranes.

    (d) Brain.

96. Which form of chromium is toxic?

    (a) Trivalent chromium.

    (b) Chromium nicotinate.

    (c) Hexavalent chromium.

    (d) Chromium in yeast.

97. Chromium enhances the effect of:

    (a) Thyroxine.

    (b) Insulin.

    (c) Theobromine.

    (d) Caffeine.

98. Chromium assists:

    (a) With blood sugar regulation.

    (b) With vision.

    (c) With kidney function.

    (d) With thyroid function.

99. Cobalt is found in:

(a) Vitamin A.

(b) Vitamin B$_{12}$.

(c) Vitamin C.

(d) Vitamin K.

100. Lead:

(a) Is an essential nutrient.

(b) Is nutritious, but nonessential.

(c) Is very toxic.

(d) Is good for children.

Appendices

Appendix A:
Pregnancy and Breastfeeding

The times before, during, and after pregnancy are of paramount importance for obtaining adequate amounts of vitamins and minerals. Excellent nutrition is also vital for breastfeeding women and young children.

All of the B vitamins are needed in greater quantities during pregnancy. Higher amounts of folic acid are needed in the diets of pregnant women to prevent neural tube defects in their children. Pregnant women are advised to take 400 to 800 mcg of folic acid to supplement the amount of folic acid available from fruits, vegetables, nuts, legumes, and in fortified food. Folic acid supplements are especially valuable from one month before conception through the first trimester of pregnancy.

Vitamin A in the forms found in animal products and in the forms found in many supplements can cause birth defects in the children of pregnant women if consumed in excess. Beta-carotene, found in fruits and vegetables, is a safe source of vitamin A for pregnant women. Pregnant women should limit their supplemental vitamin A to 5000 IU or less daily.

Abundant vitamin K in the diet is important for breastfeeding women, especially in the first few weeks of breastfeeding. This vitamin helps reduce excess bleeding in the newborn. Supplemental vitamin K is normally prescribed for newborn infants. Vitamin K is abundant in green vegetables.

Calcium and phosphorus are needed by pregnant women for fetal bone formation. Phosphorus is needed to form the structure of both DNA and RNA. Adequate dietary calcium is needed to prevent calcium losses from bones of the mother.

During pregnancy, if insufficient dietary calcium is consumed, calcium and lead could be removed from the bones of the mother and put into the blood circulation with damaging effects on the fetus. Lead can easily pass through the placenta and can damage the vulnerable nervous system of the growing fetus. Furthermore, withdrawal of calcium from the mother can increase the risk of osteoporosis later in her life.

Iron deficiency is common in pregnant women. In the last six weeks of pregnancy, women need more iron than even the best diet can provide. Especially during these weeks, iron supplementation is recommended so that the iron stores of the pregnant woman are not depleted.

Because extra iron supplementation can interfere with the absorption of zinc from foods, zinc supplements may be needed along with iron supplements. Zinc is especially needed during pregnancy and for the first six months of breastfeeding. Four out of five pregnant women worldwide have inadequate zinc levels. Adequate zinc is associated with healthier babies and easier deliveries.

Iodine deficiency can cause brain damage, especially when it occurs in the fetuses of pregnant women after the first trimester and in children of up to three years of age. Iodine is critical for the growth and development of the brain and central nervous system. Stunted growth, mental retardation, and cretinism can affect the children of pregnant women with inadequate iodine intake. Pregnant and breastfeeding women can ensure adequate iodine by taking a daily prenatal supplement providing 150 mcg of iodine.

Pregnant women sometimes suffer from water retention. Lowering sodium amounts in the diet to recommended levels may help reduce swelling.

To ensure optimal health for both the baby and the mother, these vitamins and minerals need to be maintained at their ideal levels.

Appendix B: Weight Loss

Weight gain can be understood as resulting from taking in more calories than are burned for energy. Limiting calories in the diet and increasing exercise are both valid approaches to weight loss. It is important to note that the ability of the body to burn nutrients for energy may be limited by deficiencies of certain vitamins and minerals.

Much of the modern diet is made from refined grains such as white flour. White flour is commonly found in breads and noodles. White flour is routinely enriched with certain nutrients. Other nutrients are reduced in the refining of the grains, and they are not added back in the enrichment process. Enrichment of white flour normally consists of adding thiamin, riboflavin, niacin, folic acid, and iron to the refined grains. White rice is also depleted of some vitamins and minerals, compared to brown rice. Niacin, vitamin B_6, and folic acid are greatly reduced during the processing of white rice.

Pantothenic acid and magnesium are two nutrients important for weight loss that are depleted during refining, but not added back to refined grains. Empty foods such as sugar and alcohol are devoid of the vitamins and minerals needed for their metabolism.

The ability to burn stored fat in the body is important for losing weight. Riboflavin, niacin, biotin, and vitamin B_{12} are all needed to prepare stored fats for burning in the metabolic machinery of the body. Even if these vitamins are present in sufficient amounts, the fats must be synthesized with acetyl-coenzyme A (also known as acetyl-CoA) in order to be burned. Pantothenic acid is a necessary component of acetyl-coenzyme A.

About 43 percent of the pantothenic acid in whole wheat is lost in the milling process. Enriched grains such as white flour are not enriched with pantothenic acid. In addition, freezing and processing can decrease the remaining pantothenic acid by approximately half. Deficiency of pantothenic acid may be one limiting factor in energy production. Pantothenic acid, in acetyl-coenzyme A, is at the center of energy production. It is needed not only for fat burning, but also in the burning of carbohydrates and protein.

Thiamin, vitamin B_1, plays a key role in the burning of energy in all cells. Thiamin is part of the coenzyme thiaminpyrophosphate. This enzyme helps convert carbohydrates to acetyl-coenzyme A. This is a normal step in the production of energy from carbohydrates. Thiamin is added to enriched grains and is not normally deficient, although marginal thiamin deficiency affects about one-quarter of the people in the United States and Canada.

Magnesium is needed to convert thiamin to its active enzyme form, thiaminpyrophosphate. Thus, a deficiency of magnesium can limit energy production in the body by limiting the amount of thiamin that can be converted to its active form. If carbohydrates cannot be burned for energy, they may be stored as body fat. An average of 76 percent of the magnesium in whole wheat is removed in the refining process and none is added back. About 70 percent of the magnesium in brown rice is lost in the conversion to white rice.

In addition to its role in activating thiamin, magnesium is needed by an enzyme that controls the first four steps of aerobic energy production. Deficiency of magnesium can slow down this energy-producing cycle. Most energy production and energy transfer in cells uses adenosine triphosphate (ATP), which exists as a complex with magnesium.

Vitamin B_6 is needed to release the energy from certain carbohydrates. Vitamin B_6 is needed by a coenzyme that helps release blood sugar from glycogen (stored blood sugar). Only 13 percent of the Vitamin B_6 in whole wheat remains after it is refined into white flour. White rice contains only 34 percent of the vitamin B_6 present in brown rice. Vitamin B_6 is not normally added back to grains during enrichment. Unfortunately, the vitamin B_6 content of other common carbohydrates, such as sugar, high fructose corn syrup, and alcohol is very low. Without enough vitamin B_6, it is more difficult to burn carbohydrates.

Energy metabolism in cells is controlled by thyroid hormones. Both iodine and selenium are needed by thyroid hormones to stimulate metabolism. Four ions of dietary iodine are incorporated into the thyroid hormone thyroxine (T4). Selenium is needed for thyroxine to be converted into the active thyroid hormone T3. Without both iodine and selenium, the thyroid gland cannot regulate metabolism. Iodine is added to salt in the United States and Canada. Even so, about 11 percent of Americans are low in dietary iodine, and about seven percent of pregnant women are deficient in dietary iodine.

Iron deficiency is probably the most common nutrient deficiency in the United States and the world. Iron deficiency affects about one billion people worldwide.

Iron is important in aerobic energy production in the cell in several different ways. Iron-containing cytochromes are essential as part of the electron transport chain. The electron transport chain moves electrons to charge up adenosine triphosphate (ATP), the energy battery of the cell. An iron-sulfur protein is also used in the electron transport chain. Another enzyme used in aerobic energy production needs iron and vitamin B_2. Iron works with vitamin B_3, niacin, to transport electrons in the electron transport chain. The final step in the electron transport chain uses a complex containing iron and two copper atoms.

Copper is used in an enzyme called cytochrome c oxidase in the energy-producing mitochondria in the cell. The availability of copper is needed to make ATP.

Potassium is needed by an enzyme called pyruvate kinase. This enzyme is used to break down carbohydrates for energy production in the cell. Without enough potassium, it may be more difficult to burn carbohydrates.

Nickel is needed by certain enzymes used in anaerobic energy production in the cell. Nickel works with iron and sulfur to release energy from carbohydrates.

The vitamins and minerals that are needed for weight loss that may be deficient in typical diets include pantothenic acid, magnesium, vitamin B_6, iodine, selenium, iron, copper, and nickel. If we are deficient in any one of these vitamins or minerals, we may not be able to efficiently burn nutrients for energy. The inability to burn carbohydrates leads to more stored fat. The inability to burn fat can contribute to excess body fat.

Appendix C: Antioxidants

Antioxidants neutralize the damaging effects of free radicals. Free radicals are molecules with unpaired electrons that steal electrons from stable molecules. The stable molecules can then become free radicals, causing a chain reaction of damage. Free radicals can damage artery walls and low density lipoproteins (LDL), increasing the risk of cardiovascular disease. Free radicals can also damage DNA, increasing the risk of cancer. A certain amount of free radicals is formed as unwanted byproducts of normal metabolism. Free radicals are also formed when hard radiation impacts living tissue. Pollution and cigarette smoke are two more causes of free radicals.

Antioxidants can be found in foods. Also, the body produces antioxidants. The antioxidants produced by the body are dependent on certain nutrients to function.

Vitamin C is one of the most powerful antioxidants. Vitamin C is found in fruits and vegetables. Like all good antioxidants, vitamin C can neutralize a free radical without becoming a free radical itself. Vitamin C can protect many indispensable molecules in the body, such as proteins, fats, carbohydrates, and nucleic acids (DNA and RNA), from damage by free radicals. The protection of DNA from oxidative damage is one way that vitamin C can help reduce the risk of cancer.

Vitamin C also has a role in recharging vitamin E and beta-carotene after they have performed their antioxidant functions. Iron is absorbed in the intestines with the help of vitamin C. Vitamin C assists the absorption of iron by protecting the iron from oxidation. If supplemental vitamin C is needed, the ascorbated form is easiest on the digestive system.

Vitamin E is a family of related antioxidants. Vitamin E is plentiful in raw nuts and seeds, especially sunflower seeds. Alpha-tocopherol is the main form of vitamin E. There are four tocopherols with proven antioxidant activity. There are also four tocotrienols in the vitamin E family with even stronger antioxidant activity. These natural forms of vitamin E neutralize free radicals in the fatty areas of the body. Vitamin E reduces the risk of cardiovascular disease by protecting artery walls. Vitamin E is vitally important for the protection of the cell membranes from free radical damage. Unfortunately, only one of the eight forms of synthetic alpha-tocopherol is in a form that is found in nature. At least half of the forms of synthetic alpha-tocopherol are ineffective as antioxidants. If supplemental vitamin E is needed, the natural form (d-alpha-tocopherol) with mixed tocopherols should have the best antioxidant activity.

Beta-carotene is one of the most powerful antioxidants in food. No other form of vitamin A has antioxidant activity. Beta-carotene is plentiful in yellow and orange vegetables and fruits. Green vegetables also are rich in beta-carotene. All of the many carotenes in food have antioxidant activity.

One of the most powerful free radicals in the body is superoxide. Superoxide is neutralized with an enzyme made in the body that is called superoxide dismutase. Superoxide dismutase converts superoxide to hydrogen peroxide, a less dangerous free radical. There are three types of superoxide dismutase. Two types of superoxide dismutase use zinc for structural stability and copper for catalytic activity. These two types of superoxide dismutase protect many areas including the red blood cells and the lungs. The third type of superoxide dismutase is found in the mitochondria and has manganese as its center. The body must have sufficient amounts of copper, zinc, and manganese for these important antioxidant systems to function.

An important free radical in the body is hydrogen peroxide. The body has an enzyme designed for the neutralization of hydrogen peroxide. This enzyme is called glutathione peroxidase. Selenium is part of this important antioxidant. Glutathione peroxidase protects against oxidation inside cells, in cell membranes with vitamin E, in blood plasma, in sperm, and in the intestines. Glutathione peroxidase has the ability to transform harmful antioxidants such as hydrogen peroxide into water. Glutathione peroxidase is then recharged with the assistance of the B vitamins riboflavin and niacin. The body must have sufficient amounts of selenium, riboflavin, and niacin to support the antioxidant activity of glutathione peroxidase.

Selenium is also found in another antioxidant, selenoprotein P. Selenoprotein P is capable of neutralizing nitrogen free radicals in the lining of blood vessels. Another mineral that is a necessary component of antioxidant enzymes is iron.

These antioxidants can slow aging and reduce the risk of some chronic diseases.

Appendix D: The Elderly

Special consideration may be needed for older adults to achieve needed vitamin and mineral nutrition. A broad spectrum of nutrients is needed to increase disease resistance. Some of these nutrients are needed in greater amounts by older adults. Certain of these nutrients need to be limited because of the special vulnerabilities of older adults.

In older men and women, long-term intakes of preformed vitamin A (forms other than beta-carotene) can be associated with increased risk of osteoporosis. Levels of only 5000 IU (1,500 mcg) are enough to increase this risk. This is well below the normal tolerable upper intake level for adults, which is set at 10,000 IU (3000 mcg) of vitamin A per day.

Older men and women may want to limit their supplemental vitamin A intake or take only the beta-carotene form of vitamin A. However, low levels of vitamin A can also increase risk of osteoporosis. In older people, abundant beta-carotene from colorful fruits and vegetables is safe and contributes to healthy bone density.

More of certain B vitamins may be needed by older people. Elderly people may be at risk of thiamin deficiency because of low intakes and reduced absorption. Folate is important for reducing blood homocysteine levels. Because older people tend to have higher homocysteine levels, they are encouraged to meet or exceed the RDA for the B vitamin folate.

The elderly are less able to synthesize vitamin D in the skin, so they need a little more sun than younger people. Also, many older people use sunscreen and wear protective clothing, which limits vitamin D production. Older adults also have a higher need for vitamin D. Adults aged 51 to 70 have an adequate daily intake (AI) of 400 IU (10 mcg) of vitamin D, but for ages over 70 the AI is 600 IU (15 mcg).

Salt sensitivity has been reported to be more common in the elderly. For the elderly, lowering salt intake to the recommended levels is especially important. The tolerable upper intake level for sodium is set at 3.8 grams for adults. For those over age 70 the upper level is set a little lower at 3 grams daily. Keeping salt intake below this limit will help older people control blood pressure and decrease their risk of heart disease.

In elderly populations absorption of magnesium may be lower. Also, magnesium losses in urine increase in older people. These factors, coupled with lower intakes, increase the risk of magnesium depletion in the elderly. Adequate magnesium is needed for energy and to lower the risk of cramps and spasms.

Older adults have zinc intakes that tend to be lower than the RDA. To avoid impaired immune system functioning, older adults should be sure to maintain adequate zinc intake. There are some indications that adequate zinc intake may reduce the risk of macular degeneration, common in older populations.

Appendix E: Alcoholism

Alcoholics have special needs for vitamins and minerals. Alcoholics often get smaller amounts of nutrients because they consume less food. Many alcoholic beverages can be seen as "empty calories" because they supply energy without the nutrients needed to burn that energy.

Alcoholics may be more sensitive to excesses of certain nutrients. People with a history of liver disease or alcoholism may be more susceptible to the adverse effects of excessive niacin intake. For those with alcoholic cirrhosis, the safe dose of iron may be lower than the normal upper intake level of 45 mg daily. Alcoholics may also be susceptible to vitamin A toxicity at low doses. This applies to all forms of vitamin A except beta-carotene.

Several B vitamins are adversely impacted by alcohol consumption. These B vitamins are essential for energy production. Thiamin deficiency is common among alcoholics. Alcohol impairs the absorption of thiamin, while increasing its elimination. Niacin is used up in the conversion and detoxification of alcohol. Also, vitamin B_{12} absorption is diminished in alcoholics.

Another B vitamin, vitamin B_6, is depleted by the consumption of alcohol. Alcohol is broken down to acetaldehyde in the body. Acetaldehyde breaks the vitamin B_6 coenzymes loose from their enzymes and the vitamin B_6 is lost.

Drinking alcohol can decrease the absorption of folate. In addition, heavy consumption of alcohol can prevent the liver from retaining folate. Alcohol increases losses of folate. Without enough folate, homocysteine levels in the blood can rise, contributing to heart disease.

Vitamin C levels are depleted by alcoholism. In addition, the alcoholic liver has problems activating vitamin D.

Magnesium depletion is frequently encountered in chronic alcoholics, both from low dietary intake and increased urinary losses. Some alcohol withdrawal symptoms such as delirium tremens may be related to magnesium deficiency. In cases of severe alcoholism, low magnesium levels can cause bone loss.

Low levels of phosphorus are rarely seen except in near starvation or in alcoholism. Alcoholism also increases the risk of low blood potassium.

Alcoholics are at increased risk of zinc deficiency both from impaired absorption and increased urinary losses. One-third to one-half of alcoholics have been found to have low zinc levels.

Keeping alcohol consumption to moderate levels will limit the health-depleting effects. Moderate levels are often defined as one drink per day for women and two drinks per day for men. Drinking alcoholic beverages with food and water can also reduce some of the deleterious effects. Special supplementation may be warranted for alcoholics who do not eat enough nutrients.

Appendix F: Osteoporosis

Osteoporosis is a disease that involves loss of bone strength leading to an increased risk of fracture. Osteoporosis afflicts 25 million people in the United States. Women are six times more likely to have osteoporosis than men. There are no symptoms unless a bone is broken.

Osteoporosis is a long-term chronic disease that normally takes decades to develop. Proper nutrition is one of the important factors in preventing osteoporosis. Supplemental calcium cannot prevent or heal osteoporosis by itself. It is interesting to note that widely varying levels of calcium intake around the world are not associated with corresponding rates of osteoporosis.

It is important to maintain flexibility and balance to reduce falls. In addition, weight-bearing exercise increases bone density. Certain vitamins and minerals play critical roles in bone mineralization and strength.

In older men and women, long-term intakes of preformed vitamin A can be associated with increased risk of osteoporosis. Preformed vitamin A includes all forms except the beta-carotene form. Levels of only 5000 IU (1,500 mcg) are enough to increase risk. This is well below the tolerable upper intake level, which is set at 10,000 IU (3000 mcg) per day. In contrast, intake of the beta-carotene form of vitamin A is not associated with any increased risk of osteoporosis.

Older men and women may want to limit their supplemental vitamin A intake or take only the beta-carotene form of vitamin A; fresh fruits and vegetables are good sources. Too little vitamin A can also be a problem because adequate vitamin A is needed to prevent osteoporosis.

Vitamin D deficiency can be a contributor to osteoporosis. Without enough vitamin D, the bones cannot properly mineralize. Either sufficient sunlight or 600 to 700 IU of supplemental vitamin D daily have been found to lower the chances of osteoporotic fracture.

Vitamin K is needed to bind minerals to bones. Vitamin K is used as a coenzyme to enable bone mineralization. Several studies have found a correlation between higher vitamin K levels and lowered risk of hip fracture.

If the amount of potassium-rich fruits and vegetables eaten is not sufficient to produce enough alkalinity to buffer blood acids, the body has the ability to remove calcium from bones. This calcium helps neutralize the blood, but leaves the bones depleted in calcium. This increases the risk of osteoporosis.

Excess dietary sodium can impact calcium levels in the body. Americans eat an average of five grams of excess sodium each day (above the upper intake levels). About 287 mg of extra dietary calcium is needed each day to make up for the calcium lost from this average amount of excess sodium. If the extra calcium is not available in the diet, then it may need to be removed from the bones—increasing the risk of osteoporosis.

Excess protein consumption can also be a risk factor for osteoporosis. On the average, American women take in 24 extra grams of protein per day. This extra protein results in an extra need for dietary calcium estimated at 140 mg per day. The consumption of excess protein is thought to increase the amount of acids that must be neutralized by blood buffers. Calcium can be depleted when it is used to neutralize the acids in blood that result from burning excess protein. If the extra calcium is not available in the diet, then it may need to be removed from the bones. This increases the risk of osteoporosis.

High levels of calcium intake and high levels of vitamin D are helpful in slowing the progression of bone loss. Loss of calcium from the bones, as can occur on a day with lower calcium intake coupled with high sodium and protein intake, is hard to replace. Calcium can be removed quickly from bones, but it is a slower process to rebuild bones. As is true in many chronic diseases, prevention is the best approach.

Magnesium is part of the structure of bones. Most of the magnesium in the body, about 60 percent, is used as part of the structure of the bones. Magnesium is needed for proper bone mineralization along with calcium and phosphorus. Parathyroid hormone and calcitriol (the active vitamin D hormone) both depend upon magnesium for the mineralization of bone. When magnesium is low in the blood, blood calcium levels can fall. Low levels of magnesium in bone cause a resistance to parathyroid hormone, resulting in less bone mineralization. Even lower levels of magnesium in bone lead to bone crystals that are larger and more brittle. This is why adequate magnesium may be a factor in preventing osteoporosis.

To sum it up, bone growth and regulation is assisted by vitamin A, vitamin C, vitamin K, calcium, potassium, phosphorus, iron, and magnesium.

Appendix G: Quick Summaries

Summary for Thiamin—Vitamin B$_1$

Main function: energy metabolism.
RDA: men, 1.2 mg; women, 1.1 mg.
No toxicity reported, no upper intake level set.
Deficiency disease: beriberi.
Healthy food sources: whole grains, and found in most raw or lightly cooked foods.
Degradation: easily destroyed by heat.
Coenzyme forms: thiaminpyrophosphate (TPP), thiamin triphosphate (TTP).

Summary for Riboflavin—Vitamin B$_2$

Main function: energy metabolism.
RDA: men, 1.3 mg; women, 1.1 mg.
No toxicity reported, no upper intake level set.
Deficiency condition: ariboflavinosis.
Healthy food sources: whole grains and green leafy vegetables.
Degradation: easily destroyed by light, especially ultraviolet light.
Coenzyme forms: flavin adenine dinucleotide (FAD), and
  flavin mononucleotide (FMN).

Summary for Niacin—Vitamin B$_3$

Main function: energy metabolism.
RDA: men, 16 mg; women, 14 mg (niacin equivalent).
No toxicity has been reported from food; flushing has been reported above
  35 mg in the nicotinic acid form; no effects have been noted below 2000 mg
  of niacinamide.
Deficiency disease: pellagra.
Healthy food sources: whole grains and nuts.
Degradation: heat resistant.
Coenzyme forms: nicotinamide adenine dinucleotide (NAD), nicotinamide adenine
  dinucleotide phosphate (NADP).

Summary for Biotin

Main function: energy metabolism.
Adequate Intake: men and women, 30 mcg.
No toxicity reported.
Deficiency is very rare.
Healthy food sources: found in a wide variety of foods.

Degradation: uncooked egg whites bind biotin.

Coenzyme forms: acetyl-CoA carboxylase, pyruvate carboxylase, methylcrotonyl-CoA carboxylase, and propionyl-CoA carboxylase.

Summary for Pantothenic Acid—Vitamin B$_5$

Main function: energy metabolism.

Adequate Intake level: men and women, 6 mg.

No toxicity or deficiency disease reported, no upper intake level set.

Healthy food sources: avocado, sunflower seeds, and sweet potatoes.

Degradation: easily destroyed by freezing, canning, and refining.

Coenzyme form: pantothenic acid forms a part of coenzyme A.

Summary for Pyridoxine—Vitamin B$_6$

Main function: building amino acids and fatty acids.

RDA: men, 1.7 mg; women, 1.5 mg.

Tolerable upper intake level is set at 100 mg daily.

No toxicity reported from food or below 200 mg per day.

Healthy food sources: bananas, potatoes, and spinach.

Degradation: easily destroyed by heat and can be leached out into cooking water.

Coenzyme form: pyridoxal phosphate (PLP).

Summary for Folate

Main function: synthesis of DNA and red blood cells.

RDA: adults, 400 mcg; pregnant women, 600 mcg.

Tolerable upper intake level: 1000 mcg.

No toxicity reported from food.

Deficiency disease: neural tube defects, anemia, and excess folates may mask vitamin B$_{12}$ deficiency.

Healthy food sources: leafy green vegetables and legumes.

Degradation: easily destroyed by oxygen and heat.

Coenzyme form: tetrahydrofolate (THF).

Summary for Cobalamin—Vitamin B$_{12}$

Main function: cell synthesis and red blood cells.

RDA: men and women, 2.4 mcg.

No toxicity reported from food or supplements.

Deficiency disease: pernicious anemia.

Healthy food sources: fortified cereals, nutritional yeast.

Degradation: easily destroyed by microwave cooking.

Coenzyme forms: methylcobalamin and deoxyadenosyl cobalamin.

Summary for Ascorbic Acid—Vitamin C

Main function: collagen formation and antioxidant.
RDA: 75 mg to 125 mg for adults.
No toxicity reported.
Mild digestive irritation possible over 100 mg with the ascorbic acid form.
Tolerable upper intake level is set at 2000 mg daily.
Deficiency causing scurvy is rare. Many people do not achieve the RDAs.
Deficiency disease: scurvy.
Healthy food sources: found in fresh fruits and vegetables.
Degradation: leached by cooking water, reduced by heat, light, oxygen, and food processing.

Summary for Vitamin A

Main functions: vision, skin, mucous membranes, preventing infection, and antioxidant.
RDA: adult women, 2333 IU; adult men, 3000 IU.
Toxicity is possible with supplemental forms of vitamin A. Toxicity is possible with high intakes from food. No toxicity is reported for beta-carotene in food.
Tolerable upper intake level for preformed vitamin A is set at 10,000 IU daily. This may be too high for older people and pregnant women.
Deficiency is uncommon in developed countries.
Deficiency disease: night blindness and xeropthalmia.
Healthy food sources: colored fresh fruit and vegetables.
Degradation: reduced by heat, light, and oxygen.
Food forms: provitamin A carotenoids in plants and retinyl esters in animal products.
Forms in the body: retinol, retinal, retinoic acid, retinyl palmitate, and beta-carotene.

Summary for Vitamin D

Main functions: keeps bones strong and helps regulate calcium and phosphorus.
Adequate Intake: under age 50, 200 IU; ages 51–70, 400 IU; age 70+, 600 IU.
Toxicity is possible only with supplemental forms of vitamin D and is rare.
Tolerable upper intake level is set at 2,000 IU daily.
Deficiency disease: rickets and osteomalacia (soft bones).
Healthy sources: sunshine. Supplementation with vitamin D3 may be needed.
Food sources: fortified milk, salmon, sardines, and mackerel.
Forms in the body: cholecalciferol, calcidiol, and calcitriol.

Summary for Vitamin E

Main functions: antioxidant protection of cell membranes and LDL.
RDA as RRR-alpha-tocopherol: adolescents and adults, 22.5 IU (15 mg); children, 6 IU to 16 IU (4–11 mg).

Vitamin E is nontoxic at less than 66 times the RDA.
Tolerable upper intake level of 1000 IU daily prevents excess bleeding.
Deficiency: one-third of adults get too little to prevent cardiovascular disease.
Healthy food sources: almonds, sunflower seeds, and cold-pressed oils.
Degradation: can be oxidized during food preparation or storage.
Principal forms in the body: alpha-tocopherol and gamma-tocopherol.

Summary for Vitamin K

Main functions: prevents excess bleeding and assists in bone mineralization.
Daily Recommended Intake: men, 120 mcg; women, 90 mcg.
Vitamin K is nontoxic. No tolerable upper level has been set.
Deficiency: rare in adults, may occur in newborn infants.
Food sources: green leafy vegetables are the best source.
Principal forms in the body: phylloquinone (vitamin K_1) and menaquinone (vitamin K_2)

Summary for Fluids and Electrolyte Balance

Water intake must equal water output.
The kidneys regulate blood pressure and blood volume.
Electrolytes are mineral ions that help regulate fluid balance.
Electrolytes are found in the plasma, in interstitial fluids, and inside the cells.
Electrolyte solutions are neutral with equal positive and negative charges.
Blood acidity is controlled by blood buffers, the kidneys, and the lungs.

Summary for Sodium

Main functions: maintains blood pressure and fluid balance, assists muscle con-
traction, and assists nerve impulse transmission.
Adequate Intake: for adults and children, it ranges from 1 to 1.5 grams per day.
Toxicity is rare. Excess intake can increase risk of high blood pressure.
Tolerable upper intake level is set at 3.8 g for adults. Over age 70 it is 3.0 g daily.
Deficiency is from excessive losses such as losses from excessive sweating.
Healthy sources: unprocessed fruit, vegetables, whole grains, and unsalted nuts.
Unhealthy sources: processed food often contains too much sodium.
Forms in the body: free sodium ion and bound to chloride.

Summary for Chloride

Main functions: maintains blood pressure and fluid balance.
Minimum requirement: 750 mg per day.
Toxicity is rare. Excess intake can increase risk of high blood pressure.
Deficiency is from excessive losses such as losses from vomiting.
Unhealthy sources: processed food often contains too much sodium and chloride.
Forms in the body: bound to sodium as salt or to hydrogen as hydrochloric acid.

Summary for Potassium

Main functions: maintains the cell membrane potential, is needed for energy production, and is used as an acid buffer.

Adequate Intake: 4.7 grams daily for adults.

Toxicity: not found with potassium in food. Toxicity from supplement overdose is possible.

Maximum potassium potency per tablet has been set at 99 mg for supplements.

Deficiency is from excessive losses rather than dietary lack.

Healthy sources: unprocessed fruit and vegetables.

Unhealthy sources: processed food often contains too much sodium and too little potassium.

Forms in the body: free potassium ion in cells and in the enzyme pyruvate kinase.

Summary for Calcium

Main functions: bones and teeth, muscle contraction, and blood clotting.

Adequate Intake: for adults and adolescents, 1000 to 1300 mg per day.

Toxicity is rare.

Tolerable upper intake level is set at 2500 mg for ages one and above.

Deficiency can cause stunted growth in children and osteoporosis in adults.

Sources: dairy products, almonds, tofu, sesame seeds, vegetables, and beans.

Forms in the body: free calcium ions in blood and hydroxyapatite in bones.

Summary for Phosphorus

Main functions: strengthens bones, in cell membranes, in DNA, maintains acid-base balance, and needed in energy transfer.

Adequate Intake: for adults and adolescents, 700 mg per day.

Toxicity is rare.

Tolerable upper intake level is set at 3 to 4 grams for ages one and above.

Deficiency is very rare.

Sources: adequate amounts are found in all food.

Forms in the body: as free phosphorus ions in blood, in phospholipids, and in hydroxyapatite in bones.

Summary for Magnesium

Main functions: strengthens bones, promotes muscle relaxation, and used to stabilize ATP.

Adequate Intake: adults, 300 mg to 420 mg per day.

Toxicity is possible from supplement use and may result in diarrhea.

Tolerable upper intake level is set at 350 mg for supplemental doses.

Deficiency is very rare. Intakes below the RDAs are common.

Sources: whole grains, nuts, leafy greens, and seeds.

Forms in the body: in bones, bound to protein, attached to ATP, and as ionized magnesium (Mg^{++}).

Summary for Sulfur

Main functions: needed for the antioxidant glutathione, part of coenzyme A, and part of the methyl donor SAMe.

Adequate Intake: none established.

Toxicity is not known.

Tolerable upper intake level has not been set.

Deficiency: not known.

Sources: food with sulfur-containing amino acids (methionine and cysteine), onions, garlic, cabbage, and Brussels sprouts.

Forms in the body: found in glutathione, coenzyme A, methionine, cysteine, and SAMe.

Summary for Iron

Main functions: transportation of oxygen and energy metabolism.

RDA: adults, 7 to 18 mg; pregnant women, 27 mg.

Toxicity: only with a genetic disorder. Digestive discomfort may occur if taken on an empty stomach.

Tolerable upper intake level is set at 45 mg. 40 mg for ages under 14.

Deficiency can cause fatigue and anemia. Deficiency is common.

Sources: meat, fish, poultry, green leafy vegetables, dried fruit, blackstrap molasses, nuts and seeds, and fortified grains.

Forms in the body: ferritin, transferrin, hemosiderin.

Summary for Zinc

Main functions: in enzymes, with hormones, protein structure, and diverse functions.

RDA: 2 to 6 mg for children and 8 to 13 mg for adolescents and adults.

Toxicity: excesses may induce a copper deficiency and gastrointestinal disturbances.

Tolerable upper intake level is 3 mg for infants ranging up to 40 mg for adults.

Deficiency can cause growth retardation and susceptibility to infection in children.

Sources: meat, fish, poultry, whole grains, and Brazil nuts.

Forms in the body: found in zinc-copper superoxide dismutase. Can be bound to metallothionein or albumin.

Summary for Iodine

Main functions: thyroid functions.
RDA: adults, 150 mcg.
Toxicity: low toxicity, excesses may cause a rise in thyroid stimulating hormone.
Tolerable upper intake level is 1100 mcg for adults.
Deficiency can cause goiter, cretinism, and brain damage to fetuses.
Sources: iodized salt, seaweed, fish.
Forms in the body: thyroxine (T4) and the more active form, triiodothyronine (T3).

Summary for Selenium

Main functions: needed for the important antioxidant glutathione peroxidase. Needed to convert thyroxine to the more active thyroid hormone.
RDA: adults, 55 mcg.
Toxicity: toxic in large amounts with skin and digestive symptoms.
Tolerable upper intake level is 400 mcg for adults.
Deficiency can reduce antioxidant activity and thyroid function.
Sources: Brazil nuts, seafood, whole grains.
Forms in the body: the major form is part of the enzyme glutathione peroxidase.

Summary for Copper

Main functions: energy production, collagen synthesis, iron transport, and as an antioxidant.
RDA: adults, 900 mcg.
Toxicity: rare.
Tolerable upper intake level is 10 mg for adults.
Deficiency can cause anemia.
Sources: nuts and seeds, avocados, and green leafy vegetables such as spinach.
Forms in the body: found in cytochrome c oxidase, lysyl oxidase, and some forms of superoxide dismutase.

Summary for Manganese

Main functions: glucose synthesis, ammonia detoxification, as an antioxidant, and in wound healing.
Adequate Intakes: adult men, 2.3 mg; adult women, 1.8 mg.
Toxicity: rare, no toxicity from food sources.
Tolerable upper intake level is 11 mg for adults.
Deficiency has few symptoms, may impair growth.
Sources: whole grains, green leafy vegetables, and peanut butter are rich sources of manganese.
Forms in the body: found in enzymes including manganese superoxide dismutase.

Summary for Fluoride

Main functions: strengthens tooth enamel.
Adequate Intakes: adult men, 3.8 mg; adult women, 3.1 mg.
Toxicity: very toxic.
Tolerable upper intake level is 10 mg for adults.
Deficiency may increase risk of dental decay.
Sources: fluoridated drinking water.
Forms in the body: fluoroapatite in bones and teeth.

Summary for Chromium

Main function: assists insulin in controlling blood sugar.
Adequate Intakes: adults, 20 to 35 mcg.
Toxicity: no reports of toxicity from trivalent chromium. Hexavalent chromium is
   highly toxic.
Tolerable upper intake level has not been set.
Deficiency may interfere with the control of blood sugar.
Sources: whole grain products, broccoli, green beans, grape juice, and spices.
Form in the body: trivalent chromium.

Summary for Molybdenum

Main function: metabolism of sulfur-containing amino acids.
RDA: adults, 45 mcg.
Toxicity: low toxicity.
Tolerable upper intake level has been set at 2 mg.
Deficiency is rare.
Sources: beans, lentils, whole grains, and nuts.
Forms in the body: found in the enzymes sulfite oxidase, xanthine oxidase, and
   aldehyde oxidase.

Answers to Quizzes

Chapter 1

1. d	2. b	3. c	4. c	5. d
6. b	7. a	8. d	9. b	10. c

Chapter 2

1. c	2. d	3. d	4. a	5. b
6. c	7. c	8. a	9. c	10. d

Chapter 3

1. c	2. b	3. d	4. d	5. a
6. c	7. b	8. d	9. b	10. a

Chapter 4

1. d	2. c	3. a	4. b	5. d
6. b	7. d	8. a	9. b	10. b

Chapter 5

1. c	2. b	3. a	4. d	5. b
6. a	7. c	8. d	9. d	10. a

Chapter 6

1. b	2. d	3. a	4. d	5. c
6. d	7. b	8. a	9. c	10. d

Chapter 7

1. b	2. c	3. d	4. a	5. c
6. b	7. a	8. d	9. c	10. d

Chapter 8

1. c	2. d	3. d	4. a	5. c
6. c	7. a	8. b	9. d	10. a

Chapter 9

1. b	2. c	3. d	4. b	5. a
6. c	7. d	8. a	9. b	10. c

Chapter 10

1. d	2. c	3. a	4. b	5. c
6. a	7. a	8. c	9. b	10. d

Chapter 11

1. d	2. b	3. c	4. a	5. a
6. d	7. c	8. b	9. c	10. a

Chapter 12

1. c	2. a	3. b	4. d	5. a
6. b	7. c	8. d	9. a	10. c

Chapter 13

1. a	2. d	3. b	4. a	5. c
6. d	7. a	8. b	9. b	10. d

Answers to Tests and Final Exam

Part 1 Test

1. b	2. b	3. c	4. d	5. a
6. a	7. b	8. d	9. a	10. d
11. a	12. c	13. a	14. b	15. d
16. d	17. a	18. b	19. c	20. b
21. d	22. d	23. a	24. c	25. b

Part 2 Test

1. c	2. a	3. c	4. a	5. a
6. c	7. b	8. c	9. a	10. b
11. b	12. d	13. c	14. b	15. a
16. b	17. c	18. a	19. a	20. c
21. a	22. c	23. a	24. d	25. b

Part 3 Test

1. c	2. d	3. d	4. a	5. b
6. a	7. c	8. b	9. d	10. a
11. d	12. a	13. b	14. a	15. c
16. b	17. d	18. a	19. b	20. c
21. d	22. a	23. c	24. b	25. d

Part 4 Test

1. d	2. c	3. d	4. a	5. b
6. d	7. a	8. c	9. c	10. a
11. b	12. c	13. b	14. a	15. d
16. c	17. c	18. c	19. a	20. b
21. c	22. b	23. a	24. d	25. d

Final Exam

1. b	2. a	3. d	4. b	5. d
6. a	7. b	8. a	9. d	10. a
11. c	12. b	13. d	14. c	15. c
16. d	17. a	18. d	19. c	20. a
21. d	22. b	23. b	24. c	25. b
26. b	27. d	28. a	29. b	30. d
31. d	32. b	33. b	34. a	35. d
36. c	37. d	38. a	39. b	40. c
41. a	42. d	43. c	44. b	45. b
46. b	47. d	48. b	49. c	50. d
51. a	52. c	53. b	54. d	55. a
56. c	57. b	58. a	59. b	60. d
61. c	62. b	63. a	64. c	65. a
66. b	67. a	68. d	69. c	70. d
71. b	72. a	73. c	74. c	75. d
76. c	77. a	78. b	79. a	80. b
81. d	82. c	83. d	84. a	85. c
86. b	87. a	88. d	89. c	90. a
91. b	92. b	93. c	94. d	95. a
96. c	97. b	98. a	99. b	100. c

INDEX